STEAM LOCOMOTIVES TODAY WITH COLOUR PHOTOGRAPHS

THE LAST OF STEAM

COLIN GARRATT

THE SWALLOW PRESS INC.

CHICAGO

Library of Congress
Catalog Card Number: 76–43194

ISBN 0–8040–0754–3

Text and photographs
by Colin Garratt

Published by
The Swallow Press Incorporated
811 West Junior Terrace
Chicago, Illinois 60613 U.S.A.

CONTENTS

1 Antiquated Survivors

Antiquated Engines

Vintage is a relative term whatever its application, and engines which eighty years ago looked rational examples of modern achievement might today look either enthralling, amusing, or even bizarre. The Gresley A4 – supersonic power of only thirty years ago – can today look very dated, smooth streamlining or not, and much the same can be said of the equally splendid semi-streamlined West German Class 10 Pacifics of 1957.

Personally I do not believe that to western eyes any steam engine can look truly modern – so conditioned are we to the rapid advances of an entirely different visual technology. Sleek chromium-plated cars, jet planes, and diesel or electric locomotives have all served not only to date the steam engine in both its mechanical and aesthetic aspects but, furthermore, to render it more akin to a prehistoric creature rampaging across the landscape – something to be marvelled at yet scarcely to be believed. This situation has come about within little more than a decade of the steam engine's disappearance from much of the western world. Throughout the following pages I shall be taking a general look at some

of the varying types and uses of the world's surviving steam engines, so what, broadly, do we mean by the term 'vintage'?

The more immediate and instantaneously recognizable examples are the low-slung boiler along with a high chimney and dome – concepts which virtually date back to the very evolution of the steam engine itself. Frequently attendant upon such designs was a meagre cab for the engine crew and a four- or six-wheeled low-slung tender. Over the earlier years of the steam locomotive's development the demand was for light trains; accordingly, the small boilers would be pitched low for utmost stability and were thus characterized by a high chimney to throw the smoke well clear of the driver's view. But from these beginnings, the entire history of the locomotive's evolution has been dominated by the continuing and insatiable demand for heavier and faster trains. Inevitably so, since the steam locomotive not only added vital impetus to the industrial revolution all over the world but, in many instances, countries were developed around the railway with towns and industry highly dependent upon it, while over the final years of the steam epoch, economic circumstances and competing forms of transport still ensured that the same demands were imposed upon the steam locomotive designer.

To meet these demands for speed and power, engine boilers grew wider and longer, and in accordance with loading gauge restrictions, which within various limitations applied all over the world, down came the chimneys in proportion, and so the basic form of the steam engine has gradually and systematically changed. The chimney is an often underestimated focal point of a locomotive's appearance and is, if the strange metaphor be forgiven, akin to the nose on a person's face, for when someone dons a false nose he becomes almost un-recognizable, however familiar he might be under normal circumstances. Thus the outlandishness of the truly 'archaic' look might be seen the clearer in low-slung rounded engines with almost indecently extended chimneys and boiler mountings. Indeed the first 0–6–0 coal engine initially introduced by Timothy Hackworth in 1829 had big 'lanky' wheels, low boiler and a chimney no less than sixteen feet high!

A personal weakness for such proportions is coloured by my earliest memor-ies of vintage engines in the 1950s when, each lunchtime, between the expresses and heavy coal trains on the Midland main line an old Johnson 2F 0–6–0 would head southwards with half a dozen waggons of cattle bones en route from Leicester cattle market to the glue factory near Market Harborough. After the dashing elegance of the Jubilees and the way in which the Stanier 8Fs plodded around with such a modern air, the old 2F, even then, looked as if she had burst the doors of a museum and run amok on to the main line. She was all wheels and chimney – perhaps the most bewitching formula of all. Her low tender, also riding on six large wheels, was enhanced by ornately curved and cut away sideplating which matched the sweeping curves of her cab and running plate splashers. The steam engine is many things to many people, but for me one of its predominant characteristics is its very outlandishness, and the fact that, today, vintage engines look so pathetically outlandish, combined with their inevitable aura of historical mystique, makes them utterly irresistible. In fact, I have traditionally assumed that ancient engines provide the greatest lure of all, since everyone seems fascinated by them. Be this as it may, the steam engine's longevity is something which is held with respectful awe, while the fact that there

are to this day centenarian engines still in active service is eloquent testimony to the dogged persistence which so delightfully characterizes this subject, and although countries throughout the world are totally committed to motive power modernization, the veterans soldier on as if, in truth, they will never die.

Without attempting to deride the dignity of old locomotives, it must be mentioned that over the years many will have seen an extensive renewal of parts, if not an actual rebuilding. But there can be no absolute criteria since some engines, in appearance at least, remain almost unaltered, and it is absorbing to examine a ninety-year-old engine to try and ascertain the degree of mechanical and visual change made to it within a period which constitutes a large percentage of the steam epoch itself.

However, there are many cases where frames, wheels, boiler shells and tenders are original. Note for instance the chimney of the 1885-built Indonesian B50 on plate 12. Doubtless this is the original – witness the upper part now almost like a molten candle top.

Quite apart from the actual veterans themselves, there exist today numerous old designs which have been perpetuated with very few alterations of detail, and whatever minor improvements might have been made are often of a mechanical nature and thus hidden away. Such engines, though of relatively slender years, possess all the older flavouring and stamp of the original.

The ability of an engine to achieve a ripe old age depends not only on its soundness of design and proportioning but on its ability to adapt to other forms of operation once it has been superseded in its original function by more modern designs. Paradoxically, this law seems to favour many of the mid-late

7

nineteenth-century main line types, since in later years they have been able to turn a hand to secondary work on branch lines or shunting yards either on account of their light axle loading or small size, which renders them economically suitable and easy to operate. At the opposite end of the scale, the more modern and highly specialized types, such as those built specifically for high speed express work or heavy main line goods hauls, cannot adapt to lesser forms of work either because of their size and weight or again because economics would favour the use of smaller or older designs. In fact, the modern and specialized forms of motive power have been doubly vulnerable in that it was those very top duties which first succumbed to the diesel and electric locomotives. Accordingly, so many of the great twentieth-century giants have been swept into oblivion, often leaving a hard core of older designs on moderate work to bring the era to a close.

One cannot generalize too far, especially since the advent of the medium-sized mixed traffic engine combined with upgrading of track, closure of remote secondary lines and dieselization of shunting duties has tended to reverse this trend in some areas and put the remaining steam workings of a given country in the hands of modern and standardized engines. However, those evolutionary niches, be they practical or economic, still remain in many parts of the world, often providing valuable work for incredibly antiquated engines.

Such gems only serve to promote our thoughts into the history, original function, and utilitarian aspects of the steam locomotives remaining with us today and so enable us to gain a perspective upon their particular place in locomotive evolution and, I believe, equally important, on our own personal feelings about them as well.

Early Evolutionary Forms

Since this chapter contains some very early forms of locomotive, a brief comment upon them might be useful. The 0–6–0 is one of the most famous wheel arrangements of all time. It first appeared in the late 1820s and for over a century formed the basis for a highly prolific range of goods engines and in Great Britain, the land of its origin, was still being built in tender-engine form as recently as the 1940s, while tank versions were prepared for industrial use in several countries right up to the 1960s. Although frequently associated with Britain, the type has been widely employed both in tender and tank variety all over the world since it has been found to provide a satisfactory adhesion and axle weight ratio for much general work.

Other noteworthy forms appeared concurrently with the 0–6–0 especially for fast passenger train working; prevalent was the 'single', which created such arrangements as 2–2–0, 2–2–2, 4–2–0 and 4–2–2. But, as previously mentioned, the quest for power, which loomed like a stormcloud over every new design produced, demanded longer boilers and greater adhesion. Thus the 0–4–2 and 2–4–0 had well established themselves by the mid nineteenth century. The first 0–4–2 appeared for the Liverpool and Manchester Railway in 1838, while the related 2–4–0 was of the same period, and together they formed an important part in developing the world's railways, especially in Britain, Europe, India, Canada and Australia.

With the exception of America, which is discussed separately, the general graduation to 4–4–0 did not occur until towards the end of the last century;

Plate 4
Opposite: Special attentions are bestowed upon one of the last Indonesian brass and copper capped Class D15 0–8–0Ts with the Klien Linder axle system. Hanomag, built in 1931 though based on a much older design. The engine is seen at Cepu in northern Java

8

the 4–4–2 Atlantic and 4–6–0 were logical extensions established by the early twentieth century. The 4–4–0 and 4–6–0 are in sadly dwindling evidence on world railways today, while only in Moçambique can the very last Atlantics be found at work.

Antiquated Engines Today

The devastating technological changes which have been wrought upon the world's transport systems over the last thirty years offer little claim to the perpetuation of such pristine simplicities as the eight per cent thermally efficient steam engine. The complete annihilation of steam from the vast North American continent some twenty years ago set an unnerving precedent against its continuance elsewhere. Since that time, the demise of world steam traction has been acute and now many countries have dispensed with it altogether. In general terms, the perpetuation of steam traction might be related to national affluence, with the richest and more technologically advanced countries such as the U.S.A., Canada, Australia, New Zealand, Great Britain, Scandinavia and much of western Europe having turned now to more modern motive power forms, whereas the relatively poorer areas, such as Indo-China, South-east Asia, South America, eastern Europe, parts of Russia and certain emergent countries in Africa are still, to a greater or lesser degree, dependent upon steam railways. Even the most prolifically steam-operated countries such as Poland and India have not been exempt from considerable incursions by diesel and electric traction, and there is not a country in the world today with a long-term intention to continue steam. With the sole exclusion of China, new steam con-

Plate 5
In evolutionary terms the 4–4–0 very quickly established itself as an important successor to the express passenger 2–4–0s and 0–4–2 of the mid 19th century. This magnificent design, which contains more than a hint of Prussian ancestry, is the Indonesian Class B51 2-cylinder compound 4–4–0 of German and Dutch manufacture from the turn of the century

struction has now ceased entirely, apart from odd, insignificant exceptions. The fact that this decade has been dominated by threats of dwindling energy resources, especially with regard to oil, seems to have offered no deterrent, and even countries devoid of indigenous oil supplies still seem determined upon motive power modernization schemes involving extensive use of the diesel – often generously supplied on advantageous terms by the developed world!

Yet there are still countless steam engines roaming all corners of the globe, made up of not hundreds but thousands of different types, and the study of their ancestry, delineation and mechanical characteristics is still both viable and fruitful. Thrilling though all this is, the locomotive heritage which has been recently lost is little short of tragic. Only twenty years ago hundreds of different steam classes were active in Great Britain and many were not only antiquated but up to seventy, eighty or more years old: living testimonals from the past days of free railway enterprise. To a lesser extent, Spain had an almost unbelievable array of vintage engines right up to the early 1960s, particularly 0–6–0 and 0–8–0 types of late-nineteenth century design, and their overnight disappearance in a ruthless modernization plan was one of the tragedies of recent years.

Only in one small area of the world can situations compared with these be found today – on the islands of Java and Sumatra in Indonesia. Here the Indonesian State Railways have no less than eighty steam classes on their books and, apart from many dating back to the last century, some are the last surviving representatives of early locomotive types. These exist on borrowed time under the special conditions of a poverty-stricken nation, and so grave are the eco-

Plate 7
Previous pages. Left:
Another Indonesian Class
B50 2–4–0 rests between
duties at Ponorogo – note
the logs on the 4-wheeled
tender. These handsome
period pieces were first
exported in 1880 from
Sharp Stewart of
Manchester

Plate 8
Top right: The last of the
Indonesian 2–4–0s. On the
left, a splendid antique in
the shape of a Class B13
tank engine from Hanomag
in 1885, whilst on the right
is a Class B50 tender
engine of 1881. These two
oil-fired veterans are seen
ending their days at
Jatibarang in northern Java

Plate 9
Bottom right: Further
vintage is found in the
Greek metre gauge Class Z
2–6–0Ts dating back from
the 1890s. Here one is seen
lying out of use on the
southern tip of the
Peloponnesus Peninsula

nomic conditions facing the country's railways that only one or two examples of a class might be active at a time.

Some of these priceless engines are illustrated on the accompanying pages. The B50s shown on plates 3, 7, 8 and 12 are blessed with a truly archaic aura and they represent virtually the last 2–4–0 express passenger engines left today. They were built by Sharp Stewart of Manchester between 1880 and 1885 for the State Railway of the then Dutch East Indies. Some fourteen remain 'officially active', though only two have recently seen regular use and these are employed on the one train per day along the lightly laid branch line from Madiun to Slahung in Eastern Java.

Also on Java are the last surviving remnants of the 0–4–2 passenger type; these were also built in Manchester but this time by the works of Beyer Peacock. Having lain derelict amid tropical vegetation for almost thirty-five years, they represent the kind of discovery normally associated with expeditions to unearth the fossilized remains of extinct mammals – especially since these engines constitute the final remnants of the Indonesian 4′ 8½″ standard gauge which ceased service in 1942! Moving another step up the evolutionary ladder, we discover some real period piece 4–4–0s in Java, represented by the turn-of-the-century B51 class. Over forty of these were sent to the State Railway of the Dutch East Indies between 1900 and 1909 and, although principally from the works of Hanomag and Hartmann in Germany, a few were built by Werkspoor of Amsterdam. In common with their relatives the B50s, only a couple of B51s were actually active when I visited Java and these, in oil-burning form, were working the branch line from Rangkasbitung to Labuan, a small town on the north-eastern coast of the island. See plates 5 and 6.

Archaic in its entirety is the subject of plate 2, for it depicts an operation more akin to the beginning of the Industrial Revolution two centuries ago than to the present day. This little railway situated deep in the tropics of northern Sumatra carries stones taken from the bed of a deep and fast-flowing river up to the State Railway's main line. Although relatively modern in construction, the Orenstein and Koppel 0–6–0T has that air of vintage which I find so satisfying. Can you imagine this little engine backing its rake of tiny waggons alongside a wide river in heavily wooded tropical terrain? Out in midstream are native rafts from which men dive down to the river bed and emerge with huge smooth stones. About two hundred hand-sized stones are as much as one boat can hold, and when this amount is reached the boats draw up to the bank and the stones are manually transferred into the railway waggons on low wicker platters – each platter holding about ten stones. When the train is finally loaded, the little tank engine wheezes its way through the forest until an enormous rope incline is reached. Here the loaded waggons are hauled up and a corresponding rake of empties comes down. The loaded train then continues its journey on the higher level behind a sister engine, an 0–6–0T built by Ducroo and Brauns of Weesp, Holland. After trundling through banana groves in a delightfully rickety fashion it reaches the connection with the main line. Here the stones are mechanically crushed prior to being handed over to the State Railway for use as track ballast. As if the entire scene were not quaint enough, the stones were, until recently, hand crushed by gangs of men with small hammers!

Add to the foregoing the fact that Indonesia holds some of the last steam tram engines which are still employed on both city and rural roadside use, along with some 0–4–0 tender engines as well, and it becomes abundantly evident that the country holds a working steam museum unequalled today.

Although the situation in Indonesia is exceptional, other countries can contribute some highly interesting material. At the time of writing, Indonesia cannot produce any centenarians but the hinterlands of Peru and Brazil will reward the intrepid explorer with classic American 4–4–0s in excess of one hundred years old – see Chapter 5. In discussing centenarians, Turkey must be mentioned with its Class 33 0–6–0s introduced in the early 1870s. These archaic-looking outside-framed engines are almost without doubt the oldest still extant in Europe and probably the most vintage-looking engines in the world. Portugal can still sport some delightful veterans on its busy metre-gauge network around Oporto, where 1886-built 2–6–0 tank engines perform with gay abandon, while on the 5′ 6″ gauge line through the Douro Valley is found a *pièce de résistance* of early 4–6–0s in the 1910-built Henschel shown on plate 10.

The countries mentioned and engines illustrated give some idea of the remarkable variety left, notwithstanding the advanced years; there are many others, some in the most unlikely settings. I can only trust that this section has given some awareness of these living museum pieces and that the ancient steam engine, so long revered, will continue to be so with undiminished passion.

The immense interest in railways has enabled a basic documentation of the world's surviving steam locomotives to be made; a task made easier by their now greatly diminished ranks. But some notable blind spots do remain, particularly China. There is much research still to be done and doubtless many discoveries are yet to be made. Vintage, at least for me, constitutes the most joyous locomotive study of all.

Plate 10
Opposite: One natural successor to the 4–4–0 was the 4–6–0, a type well represented by this stylish and curvaceous design from Henschel in 1910 for the 5′ 6″ gauge Portuguese Railways. She is seen in the Douro Valley where these engines are still hard at work today

Plate 11
Following pages. Left: Copper cap chimneys and brass domes gleam against the sunset and highlight an Indonesian Class C12 2-cylinder compound 2–6–0. Although this engine came from Hartmann of Germany in 1895, the basic design dates back to the 1870s

Plate 12
Right: With steam leaking from the cylinders and wood sparks blowing from the chimney, an Indonesian Class B50 2–4–0 struggles against a side wind whilst working a passenger train. Note the Indonesian state flags adorning the engine to commemorate Independence Day

2 Variants by Evolution

The Basic Concept

For all the visual prominence of the old engines previously discussed, they follow a basic and conventional pattern. It is remarkable that over some one-and-a-half centuries the steam engine differed little in basic concept from that evolved by Stephenson in 1829 with the *Rocket* – actually the twenty-ninth locomotive built, but one in which the fundamental principles were laid down.

Since the locomotive's inception some one-and-three-quarter centuries ago, many thousands of different designs have been prepared, notwithstanding that a large proportion of them have been virtually identical in mechanical principle, specifications, and power output. In fact, the biggest difference in most cases was the external appearance in which, unlike the mechanical aspect, infinite logical variations are possible.

All these designs have been diligently prepared for innumerable regimes which, within themselves, have existed in a permanent state of flux, and any change in the administrative *status quo*, be it nationalization, company mergers, wars, change of chief mechanical engineer, or political squabbles, to mention but a few, has invariably given way to a plethora of new designs often little different in basic function from the earlier ones. Obviously, the mechanical improvements introduced over the years have woven a thread of logic throughout, yet inevitably these have served to extend the variety even more. Nevertheless such aspects as valves, valve gears, boiler pressures, compounding, superheating, wide fireboxes, exhaust events and Giesl chimneys are but some of the more important variations played on the basic Stephensonian concept amid its otherwise incredibly duplicated manifestations.

Where such variations might be said to represent an evolutionary departure from the basic principle is an arbitrary matter, but in this chapter I would like to consider a few obvious variants, both mechanically and in appearance, which are still highly active on our railways today.

The Mallet

'From little acorns grow the oaks', is a relevant metaphor to apply to the Mallet, since this form of locomotive, which began as a diminutive tank engine, ultimately flowered into the biggest steam locomotives of all time: the Union Pacific 4–8–8–4 'Big Boy' Mallets which in full working order weighed some 550 tons.

The type's modest beginning came in 1884 when a Frenchman, Anatole Mallet, devised a semi-articulated tank engine in which the main frames were split into two units: the rear unit was rigid whilst the leading one was articulated. Mallet, whose principal interest lay in compounding, provided his engine

Plate 13
A Franco-Crosti boilered
Class 743 2–8–0 of the
Italian State Railways. One
of the most enthralling
steam survivors left in the
world today

with four cylinders – the two high-pressure ones driving the rear fixed unit and the two low-pressure ones operating the leading articulated one. Thus by feeding the steam from the dome directly down to the high-pressure cylinders he avoided the need for flexible high-pressure steam joints.

Soon the Mallet was widely heralded for working over lightly laid and sharply curved tracks because of its relatively light axle loading for a given power output and its ability to traverse curves. Mallet's first engine was of the 0–4–4–0 type for a sixty centimetre gauge track and it was on narrow-gauge industrial lines that the type found an early prevalence, though this was quickly paralleled by its importance as a main line engine. Portugal was an early recipient, and soon after the turn of the century both 0–4–4–0 and 2–4–6–0 four-cylinder compound tank engines were ordered for the difficult mountain routes of the metre gauge lines in the Douro Valley. One is shown on plate 21; almost all of these early Portuguese Mallets survive in service today. One year after the first Portuguese Mallet was delivered in 1905, Hungary took up the type in tender-engine form for heavy main line operation and eventually became the most important main line Mallet user in Europe. With both tank and tender versions in operation the Mallet was to be the most widely used articulated locomotive of all time: about five thousand are believed to have been built.

Apart from its splendid fulfilment with the North American giants, the Mallet also had its greatest use in the United States with some three thousand five hundred being put into operation. The highly Americanized 0–6–6–0 Baldwin example on plate 25 is historical, as it was with this wheel arrangement that the Mallet first saw use in America, when one was built for the Baltimore

and Ohio Railroad in 1904.

An early fault of the breed was sluggish running because of instability and slipping on the leading low-pressure articulated unit where lack of weight and the thrust of the big diameter cylinders threw the front unit around. But design improvements in America, combined with a change to simple expansion, largely overcame the difficulties and many mammoth engines were created. In spite of the final American version the classic Mallet is still said to be a four-cylinder compound.

South Africa was a one-time user of the Mallet, but after extensive trials the later and technically superior Garratt articulated superseded it. In fact, the Mallet made North America its major habitat, whilst the Garratt colonized Africa. Exactly why the Americans never once used the Garratt is still pondered upon today, and whether or not they would have been forced to adopt it had steam development continued is equally interesting. One American criticism of the Garratt was that, as coal and water supplies became used up, a loss of adhesion occurred – a problem not applicable to the Mallet. Such criticism has hardly been borne out by an extensive use of the Garratt type, but such were the tremendous tonnages moved by the American Mallet that every pound of adhesion was doubtless valuable.

Mallets Today

The Mallet's use as a plantation engine is still prolifically apparent over the vast palm oil plantations of North Sumatra where some one hundred and fifty of the type in 0–4–4–0 tank engine form are still operative; many being built by Ducroo & Brauns of Holland when the plantations were being developed by the Dutch earlier this century. Widespread as these plantations are, the region around Saintar is especially noted for the Mallet today. On the neighbouring island of Java a good number of Mallets can also be found operating throughout the sugar fields. The plantation Mallets of these two islands are maintained in an excellent condition, doing much useful work, and they will certainly be among the very last Mallets left in the world.

Java is recognized as holding the last big main line tender Mallets left today and it is a pity that I cannot record these engines as being in fine condition as well; unfortunately the reverse is true. For some time I had been fostering dreams of seeing the giant DD52 2–8–8–0s in action, for these are little other than a scaled-down version of bigger American engines, having been directly descended from an Alco class first delivered to Java in 1916. However, all were out of action owing to a lack of shopping facilities and spares, with the result that much of the traffic over the line they worked had been cancelled until further notice: such are the tragic economic conditions besetting the Indonesian State Railway.

Mallet operations in Java today are centred around Cibatu, a small town set amidst a range of volcanic mountains. Here I did discover one of the slightly smaller CC50 2–6–6–0s – plate 84. This engine was sharing duties on the Cibatu–Cikajang branch line with a Mallet 2–6–6–0 tank engine of class CC10 – again only one of these was active, a veteran from Hartmann of Germany in 1905. Plate 18 shows the two side by side with tank and tender versions both in compound form: the two variants of classic Mallet design! The energy and tension depicted in this scene is somewhat illusory because both engines had

failed in traffic within twelve hours of this picture being made – the tank engine having completely dropped its motion on one side.

The line was promptly closed until further notice, since these were the only two engines operable at Cibatu. Thus with freight services cancelled on the main line owing to the failure of the DD52s and the Cikajang branch also stopped, there were no steam workings at all from Cibatu – that world-famous all-Mallet depot! But this was not a situation new to the Indonesians; they had seen it all before, and a combination of prayers, ingenuity and perseverance would doubtless get the wheels turning again – tomorrow, next week or perhaps next month!

In my opinion, the world's most fabulous steam survivor is the 3′ 6″ gauge Baldwin 0–6–6–0 Mallet of 1925 which works mahogany logs from the mountain stands down to the sawmills of the Insular Lumber Company on remote Negros Island in the Philippine group. She is the Philippines' only Mallet, but what the country lacks in numbers it gains in aesthetics with this incredible engine, for the dictionary simply does not contain enough superlatives to do her full justice. She is featured on plates 14 and 25 and discussed in wider context at the end of Chapter 5.

As already mentioned, another active Mallet colony is found in northern Portugal, particularly on the densely operated metre gauge passenger services radiating out of Oporto Trindade Station. These are ably handled by Henschel 0–4–4–0Ts built between 1905 and 1908. The Douro Valley main line heads eastwards from Oporto to the Spanish border and this 5′ 6″ gauge route is intercepted by various metre gauge feeder lines from the mountains. On these

Plate 16
Perhaps the most important evolutionary offshoot from the mainstream of steam locomotive development was the British-designed Garratt type. Here a Rhodesian 2–6–2 + 2–6–2 Class 14A Garratt takes water whilst heading a West Nicholson–Bulawayo goods train. The engine was exported in 1954 from Beyer Peacock of Manchester

feeders, especially the one between Regua and Chaves, the 2–4–6–0 Mallet tank built from 1911 by Henschel of Germany is found – see plate 21. The narrow gauge lines of Syria also see limited use of the type in 0–4–4–2 tank form.

With the Mallet having all but disappeared from the rest of Europe, and the ever doubtful anomalies of Russia and China, where there might still be active, though inaccessible, Mallets, it is probably Brazil which takes the next priority. Here on the Donna Thereza Christina coal railway are possibly the last simple-expansion Mallets running today. Built in 1950 by Baldwins, these are modern brutes of the 2–6–6–2 tender type and their slogging performances render them a dynamic attraction: three are active. If the visitor to South America finds these a little too modern for his palate, he may prefer an expedition into the Chilean desert where the world's last Kitson-Meyer 0–6–0 + 0–6–0T articulated is working a nitrate mine, the Kitson-Meyer being an articulated relation of the Mallet.

The Garratt

Africa remains the last major stamping ground for what is perhaps the most important evolutionary offshoot from the mainstream of steam locomotive development, and one which has created tremendous intrigue ever since its inception earlier this century: this offshoot is known as the Garratt Locomotive. When the first Garratt engine breathed fire in 1911, articulated steam was active in various forms, notably of course in the Mallet, and although the Garratt was technically superior and did supersede its rival in certain areas, the two forms developed side by side until the end of the steam era. The last Mallet

24

was constructed for Indonesia in 1961 and the last Garratt for South Africa in 1968. Unlike the Mallet, the Garratt was a four-cylinder simple almost without exception.

In many parts of the world there eventually arose the problem of how to operate heavy trains over the graded, curved and lightly laid routes which, out of economic necessity, had to be laid when countries first opened up. Double heading of smaller engines, inefficient though it was, seemed the only answer. The Garratt engine solved this problem and was in effect a locomotive engineer's dream come true. Like many great inventions its principle was remarkably simple.

The boiler and firebox on a Garratt engine can be built to whatever size is necessary because they are completely free of wheels underneath – an exciting concept not found on other steam locomotive designs. Conversely, as the wheels are free of the boiler, they too can be made to whatever diameter is considered best. Then, by abolishing the idea of a tender and bringing the coal and water supply on to the main engine via a leading water unit ahead of the boiler and a combined coal and water unit to the rear, maximum adhesion weight is provided, though ingeniously spread over a wide area with a large separation gap beneath the boiler – because the two sets of wheels and cylinders have been placed underneath each unit! By articulating these units from the boiler proper, a large, powerful locomotive can be built capable of moving heavy loads over curved, graded and lightly built lines. Lack of restriction in firebox design means that it can also be built very deeply for maximum combustion, a great advantage when consuming the poor coal so frequently

Plate 17
A teak-burning 2-truck Shay from Lima chatters its way between the sawmill and planing mill of the Insular Lumber Co, Negros, Philippines

25

encountered, while the fat boiler, being characteristically short in length, facilitates rapid steaming since the short tube lengths provide efficient heat transference: the slogging pugnacity with which Garratts undergo their duties gives good testimony to these aspects of the design.

H. W. Garratt was an English engineer who took out a patent on this principle in 1907. Shortly afterwards Beyer Peacock of Manchester took up his idea, initially in a cautious way, though later with tremendous vigour. Unfortunately Garratt died in 1913 at the age of 49 and although he lived long enough to see the first fruits of his idea exported from England, he was never to know the great contribution he had made to steam locomotive development. Thenceforth the Garratt became associated with Beyer Peacocks, the majority being built by this company, and the engines were widely known as Beyer-Garratts. Some, however, were constructed under licence from Beyer Peacocks both by British and foreign builders.

Wafted by Britain's colonial markets, Garratts spread to many parts of the world where they were eminently suited to the rugged terrain and difficult operating conditions. As many countries became increasingly industrialized, heavier demands were made upon the railways, and under such circumstances the Garratt engine was able to come ever more into its own.

Such difficult conditions hardly applied in Britain, yet many remember with awe the solitary 1925-built London & North Eastern Railway U1 2–8–0+0–8–2 Garratt – the largest steam engine Britain ever had. Even more famous were the thirty-three London Midland & Scottish Railway Garratts, rather discredited today as being a deviation from sensible policy, yet fondly remembered by all who ever saw them. No Garratt ever had a greater axle loading than the 21 tons possessed by these 2–6–0+0–6–2s.

Britain had the largest number of main line Garratts in Europe, Spain being the only other notable recipient. The largest Garratt ever built went to Russia, a country not normally associated with the type. This solitary giant with a twenty-ton axle loading was designed to haul 2,500 ton trains, but the type was not perpetuated. India, New Zealand, South America and particularly Australia were all Garratt users.

Garratts will always be associated with the African continent. From Algeria and Sudan in the north, East Africa, Angola and Rhodesia in the central area, through to Moçambique and South Africa, the Garratt has seen extensive operation. Many exciting classes have been produced for working throughout these regions, principally for slogging freight hauls, but also for passenger duties. Whether it be on the tortuous climbs inland over the steep coastal escarpments, or amid the interior where the rough landscape causes the track beds to undulate like corrugated iron, the African Garratt performs supremely.

Being a product of the twentieth century and a major variant on the conventional steam locomotive, it might appear strange that the Garratt has not seen wider use. Certainly the oft-stated theory that steam had been developed to its ultimate potential within permissible loading gauges, thus heralding a necessity for an alternate form of motive power, can easily be refuted, owing to the enormous possibilities for expansion within the Garratt concept. South Africa has done greatest justice to the type ever since its introduction there in 1919. For all the inherent possibilities, it seems ironic that less than two thousand Garratts were ever built, though perhaps the best epitaph for them is the fact that a good percentage are still in operation.

Even in 1973, as I watched the endless succession of two-thousand-ton freight trains double-headed by large 4–8–2s over the densely operated line between Kroonstad and Bloemfontein in the Orange Free State of South Africa, I could not help wondering why those trains were not operated by one suitably proportioned Garratt. Apart from the many peripheral savings, each train would need one engine crew less, while a margin for even more power than at present would clearly exist. Our conjecture is endless.

Garratts Today

For those who fondly revere memories of the main line Garratt in England, or are simply inspired by the thought of seeing big, four-cylinder articulated steam power fully unleashed, a trip to South Africa is to be greatly recommended. Countless Garratts abound throughout the Republic, ranging from the diminutive 2′ 0″ gauge engines of the Cape and Natal to older 3′ 6″ gauge 2–6–2+2–6–2 main-liners of the 1920s. The masterpieces however are the enormous GMA/GMAM class 4–8–2+2–8–4. These were the most numerous Garratt class ever built, with one hundred and fifty examples being put into traffic. Despite their size, the GMAs have an axle loading below fifteen tons and are often referred to as branch line Garratts, although their tractive effort is 68,800 lbs.

Secondary line engines or not, their performance over the tortuous lines in Natal must constitute one of the most thrilling steam sights in the world, especially when two of them, each attached to an auxiliary water tank holding 6,810 gallons, attack the one-in-thirty grades dragging one-thousand-ton freight trains behind them. The syncopated rhythms of eight cylinders issue a roar of such intensity as to shake the earth, as the mechanically stoked giants bite into the grades. The big Mallet may well have been the steam age's 'Tyrannosaurus Rex', but some of the dramas enacted by the Garratt in South Africa are not far behind in terms of sheer spectacle.

The distinction of having the greatest dependence upon Garratts must, however, go to near-by Rhodesia where one hundred and fifty of them, embracing four different classes, constitute some ninety-five per cent of the country's highly utilized steam fleet. It is still the Manchester built Garratt which takes the huge coal hauls across Rhodesia, from the Wankie Coalfield near the Zambian border, into Bulawayo. This work is partly undertaken by the ultimate in Rhodesian steam traction – the ninety-five foot long 20th class 4–8–2 + 2–8–4, built between 1954 and 1958. By 1975, Bulawayo possessed the world's largest Garratt operation, since the majority of Rhodesia's steam engines are now concentrated there. Imagine walking through the depot with some seventy enormous Garratts looming up on all sides, the tremendous aura of power, and hardly a conventional engine in sight. Such a scene could hardly have been imagined when H. W. Garratt first called upon Beyer Peacock to discuss his idea for an articulated engine nearly seventy years ago.

Dynamic though operations are in the south, many find the East African Railway Garratts the greatest attraction of all for, apart from their superb maroon livery, they also include the biggest steam engines running today. These are the 'Mountain' class engines which move with all the stealth and majesty of a lion as they ply their twelve hundred-ton trains over the historic railway from Mombasa on the Indian Ocean to Nairobi – 332 miles inland and nearly a mile above sea level! Named after the highest mountains in East Africa, their nameplates, in the best of British traditions, carry a small emblem stating the height in feet of the particular mountain commemorated. All thirty-four came from Beyer Peacock in 1955, and within their twenty-one-ton axle load they pack an 83,350 lb tractive effort – an interesting comparison with the old L.M.S. Garratts which, for an identical axle load, produced a figure little better than half of this.

Plate 22 depicts a 'Mountain' skirting the edge of Kenya's Tsavo Game Reserve at Voi. Nearby it is possible to see the world's biggest animals and steam engines watering side by side; the elephants drink from the river bed at Tsavo River, while on the adjacent embankment are situated the engines' water columns. This scene constitutes one of the finest railway scenes imaginable and is the crystallization of David Shepherd's wonderful film *The Man Who Loved Giants*.

Moçambique can also provide excellent Garratt performances, among which can be found some ex-Sudan 4–6–4 + 4–6–4s, while on the opposite side of the continent the wood-burning Garratts of Angola provide a spectacle no inveterate traveller in search of steam will want to miss.

By comparison with Africa the world distribution of Garratts today is relatively insignificant. All European examples are now withdrawn, sadly without preservation of an English main-liner, although in recompense a small industrial engine from Baddersley Colliery has been kept. Fortunately the first Garratt ever built, a 2′ 0″ gauge 0–4–0 + 0–4–0 for the Tasmanian Railways in 1909, has not only been preserved but actually returned to Britain, where she can now be

Plate 19
Opposite: The Dolomite mountains of northern Italy reverberate to the roar of an Italian Crosti-boilered Class 741 2–8–0 seen at the head of a passenger train from Fortezza to San Candido

28

Plate 20
The ultimate in main line
steam power for South
Africa are the 108 ft long
Condensing 4–8–4s built
for service across the
waterless Karroo Desert.
With the blow-down valve
gushing out impurities
from the boiler one heads
through the desert with a
northbound express from
Cape Town

seen on the Festiniog Railway in North Wales. This original engine is, by
exception, a four-cylinder compound.

Unhappily, the Garratt has completely disappeared from Australia, and even
South America, haven of many fine things, has now all but dispensed with their
services. Although Asia is not traditionally associated with the Garratt locomotive, a handful of active examples exist in both India and Burma. Far from
being a visual desecration of the steam engine's traditional form, the Garratt can
be, in its own right, an exceedingly handsome machine though undoubtedly an
acquired taste – a few weeks among them in Africa would more than bear out
the truth of this statement.

The Shay

Almost exclusively associated with America's Pacific Coast logging railways, the
Shay might be regarded as one of the rarest, most rustic and unusual of all
steam variants. This limitation in its distribution has rendered the Shay a
relatively little known type, yet almost three thousand were built – a figure well
in excess of the world-famous Garratt locomotive.

First introduced in 1880 as the brainchild of Ephraim Shay the type was, in
common with the Garratt, associated with one particular builder, in this case
the Lima Locomotive Works of Lima, Ohio. This company took out a patent on
the design and became almost entirely responsible for Shay construction. Being

30

predominantly for home use, the majority were built to the standard 4′ 8½″ gauge, though a few of narrow gauge construction were exported, notably to Formosa (Taiwan), the Philippines, Australia and South America. With the Shay's eventual extinction in its North American habitat, it is to the remnants of these few exports that the design owes its existence today.

The Shay's metabolism is geared to a specific function – the operation of heavy trains over remote backward logging railways where the tracks are not only light in construction, heavily graded and tightly curved, but invariably badly maintained, while their engulfment in a quagmire of mud would be equally characteristic. Under such conditions, running speed becomes subordinate to good articulation and an efficient transmission of power to move the train. Accordingly, Shays are flexibly mounted on four-wheeled bogies known as trucks; many are of the two truck variety in which the leading bogie is set immediately beneath the smokebox and the rear one just behind the cab. The cylinders, usually two or three in number, are set alongside each other in a vertical position just ahead of the cab on the engine's right-hand side. These cylinders drive a horizontal crankshaft running the entire length of the engine, the drive being applied by pinions slotting into bevelled gears on the truck wheels. The single crankshaft, which drives all trucks, is made flexible by incorporating universal joints placed at intervals throughout its length. The gearing ratio applied to the small wheels gives an even turning movement and so reduces slipping.

Traditionally the cylinders and drive are always situated on the engine's right-hand side with the boiler offset to the left for weight compensation; the unusual effect created by boiler displacement is well illustrated on plate 57. The truck combinations are either two or three in number, although some four-truck Shays have existed whereby the extra truck carries an additional fuel tender.

The Shay is especially characterized by its slow speed, frequently around ten miles per hour, with twenty miles per hour in the upper limits, but what it lacks in speed it gains in sheer racket. The cylinders drive the tiny wheels with a tremendous roar accompanied by a grinding cacophony of gears as the wheels bounce over uneven track-beds, the gnashing and clashing doubtless accentuated by wear in the engine's axleboxes. During my stay on Negros I listened to the Shays with blank amazement since the noise, combined with their wonderfully hideous shape, provided a totally new experience in watching steam engines. But it was at night that they really excelled themselves, when from the huge spark-arresting chimneys came flurries of brightly coloured teak embers which sprayed the surrounding area with all the velocity of an erupting roman candle – see plate 17.

Shays Today

Recent research has made the remaining Shay colony on Taiwan by far the best known today. Here, on the Ali-Shan mountain railway, both two- and three-cylinder varieties, weighing eighteen and twenty-eight tons respectively, bring cedar wood down from the mountain stands over what must be one of the most spectacularly built railways in the world – climbing 2,273 metres over a seventy-two kilometre journey from the west coast main line connection at Chia-Yi to Ali-Shan. Work for a Shay indeed! Spirals, switchbacks, tunnels, wooden trestle viaducts, sinuous curves and tortuous gradients all in evidence.

Awe-inspiring though the Ali-Shan is, the Shay engines belonging to Insular Lumber on Negros Isle, Philippines, are infinitely more exciting, though admittedly they operate over a structurally less spectacular line. These decrepit examples possess wooden buffer beams, home-made spark-arresting chimneys of unprecedented proportion and a glorious range of colour tones, including not a little rust. In fact they are a wonderful testimony to the longevity and mechanical adaptability of the design, and furthermore, to the way in which a steam engine will soldier on with both limited maintenance and adverse operating conditions.

A number of static examples survive 'stuffed and mounted' in North America.

The Franco-Crosti

Long after the steam locomotive has disappeared from the world's railways, engineers and enthusiasts alike will ponder upon how it might have been further developed and improved. Had it not been so ruthlessly swept aside, what might its future potential have been? Although many countries perpetuate steam, not one is conducting any tests or serious research; it is almost universally fashionable to regard it solely as a thing of the past.

Over the later years of steam development a number of serious investigations were carried out in an attempt to improve on the basic theme. One of the most dramatic was the Italian Crosti boiler. Over much of this century, Italy's steam fleet has been composed of a distinctive, well standardized but rather undistinguished set of engines seldom noted for any dynamic performances. Nevertheless, this group of carefully produced standards provided Italy, for better or worse, with a backbone of power which, almost seventy years later, remains operative today – Italy's last steam engine was built some fifty years ago! Since then, much has been put into acquiring alternative power forms, especially electricity, owing to Italy's abundance of hydro-electric potential, but perhaps inevitably Italian ingenuity did come up with some intriguing ideas for improving steam.

The most noted is the Crosti boiler, an innovation aimed at increasing the steam engine's low thermal efficiency and so reducing coal consumption. It is natural that ideas like this should germinate in a land devoid of coal fields, but allegedly the idea was further precipitated by a political squabble with Britain in 1936, which resulted in imported supplies of British coal being cut off.

Conceived in 1937 by Dr Piero Crosti, the principle was initially applied early in 1940 when several examples of Italy's standard freight engine, the 740 class 2–8–0 of 1911, were dramatically transformed by the addition of two secondary boilers slung either side of the main one – see plate 13. The engine's original boiler was retained, but its tube elements were altered to release fire gases into the smokebox at a higher temperature than usual, then, by feeding them back through tubes placed in the pre-heaters, the gases traversed the engine twice and were finally ejected at a very low temperature from chimneys placed either side of the engine immediately in front of the cab.

Apart from producing a remarkably grotesque variation on the basic theme, the engines concerned also have a beheaded appearance, since there is no purpose for a conventionally placed chimney. A reduction of ten per cent in coal consumption was enthusiastically reported and by 1953 eighty-nine 740s had

Plate 21
Opposite: A Portuguese metre gauge 2–4–6–0 Mallet tank engine departs from Regua with a branch train to Chaves. She was built by Henschel in 1923 to a design of some 12 years earlier

been converted, whereupon they were reclassified 743. Many standard 2–6–0s were similarly rebuilt.

Plate 19 shows a further development within this context, known as the Franco-Crosti, when a 740 was rebuilt with just one pre-heater unit, this time slung underneath the main boiler. The principle was identical, but apart from producing a more gainly looking engine, only one chimney was necessary. This was placed on the fireman's side. Over the period 1959/60, eighty such engines were converted, these being reclassified 741.

Brilliant though the principle was, it did possess some inherent defects, particularly metal corrosion in the pre-heater tubes and chimney caused by the low temperature gases releasing sulphuric acid which rotted the metal, while a weakened exhaust owing to long passages did little to promote lively steaming. However, improvements were made which indicated that, in the long term, these problems were solvable. However, to see these engines running around with jagged and half eaten-away chimneys was amusing, to say the least.

A handful of other countries toyed with the idea, namely Germany, Belgium, Spain and Great Britain, where ten of British Railway's last steam engines had Franco-Crosti boilers. These engines, like the LMS Garratts of twenty-five years earlier, were used on the big coal hauls along the ex-Midland main line. Rather delightfully, the British and German engines retained false chimneys on their smokebox tops and, though these were actually used when lighting up, they were no doubt in deference to a crumb of visual sanity as well.

To sum up, the innovation, in common with the Giesl Ejector, came too late for serious application, and the Franco-Crosti boiler will pass into history as a most gallant, and successful, attempt to improve the steam engine's efficiency.

Franco-Crostis Today

Only in the land of origin can the final remnants of this type be found and these are solely confined to the 2–8–0, all 2–6–0 variants having since disappeared. However, both the 741/3 versions are still active, but according to official sources the 743 is expected to be withdrawn before the end of steam traction in Italy. This is a shame, since these are the more interesting and already their numbers have dwindled to a precariously low level.

Along with some original and unaltered 740s, both variants work mostly in Northern Italy, where the industrial concentration and dense population provide many secondary lines over which present-day steam power operates. As plate 19 shows, 741s even head the Fortezza–San Candido passenger trains through the Dolomite Mountains, while further south their 743 relations are making Alessandria, Cremona, Cassino and Torino noted in steam locomotive history.

I have never failed to wonder at the little interest shown in these engines – even the many preservation schemes seem to have by-passed them. Two years ago the Italian State Railways announced that they had no plans to preserve one. No other country which adopted the principle has retained any examples. Although I would not decry the brilliant preservation work being carried out in Great Britain today, it does rankle a little to see so many 'conventional' engines being imported for preservation and the complete disregard of these historical 2–8–0s. Fortunately there is some time yet for this deficiency to be made good.

Condensers

A recent tragedy befalling main line steam was the removal from the Karroo Desert of the stupendous 108-foot-long South African Railways condensing 4–8–4s. Apart from being the only big condensing engines left in existence, they were, along with some non-condensing sisters, the last main line steam engines supplied to South Africa.

Having spent several days in the Karroo watching them in action, I never failed to be enthralled by the consummate ease with which they lifted the two-thousand-ton trains away from a stand, accompanied by a glorious scream of turbine fans. The sound is almost identical with that of a jet plane on take-off, and it is a remarkable experience to be standing at the trackside watching a condenser approach – a huge, thundering locomotive billowing up smoke yet sounding just like an aircraft! Condensers have no conventional exhaust beats, because after the steam has driven the pistons it passes through to condensing elements set in the tender, and the draught required for smoke emission is artificially created by a steam turbine-driven fan placed in the smokebox bottom – the exhaust steam, however, does drive the turbine.

Imagine the effect when two Condensers are coupled together on a three-thousand-ton freight train – two hundred and sixteen feet of locomotive at the head of one train! It is a grand sight – both engines belching a tumult of black smoke skywards, their seventy-square-foot grate areas being mechanically stoked, as they are beyond the physical capacity of one fireman to maintain steam, especially when hauling such tonnages.

Their design was conceived by Henschel of Germany who actually built the initial engine in 1953. The South African Railways had always experienced difficulty in getting steam locomotives across the desert, especially during the dry season when water would have to be specially transported to certain pick-up points – a costly and troublesome operation. Henschels were pioneers with condensing engines, building them for Argentina, Iraq and Russia. Also they equipped some of Hitler's 2–10–0 war engines with special condensing apparatus as part of Germany's plan to conquer Russia, plate 75.

After Henschels produced the initial South African engine, the remaining eighty-nine all came from the North British Works in Glasgow over 1953/4 and were the last important main line engines ever built by that company. The frames and cylinders are in a one-piece steel casting which came from the General Steel Castings Corporation, U.S.A. – a precedent set by the earlier 24 class 2–8–4s of 1949/50, illustrated on plate 87. Construction of the intricate condensing tenders to contract price is said to have been a serious financial embarrassment to North British.

Delivered at a cost of £112,000 each, the Condensers quickly revolutionized traffic through the arid region. Their capacity to run seven hundred miles without need to replenish water in their 4,500-gallon tanks was little short of a miracle. After driving the pistons and exhaust turbine, the steam travels along a sixteen-inch diameter pipe situated along the engine's running board (plate 24), and after passing through an oil separator, it powers another turbine which, by means of a common shaft, drives five enormous air intake fans situated in the tender top. These draw in cool air through meshing in the tender sides, and the steam is then brought into contact with condensing elements set on both sides of the tender. The condensate accumulates in a 600-gallon tank before being

Plate 22
Following pages: These magnificent 'Mountain' Class 4–8–2+2–8–4 Garratts of the East African Railways are believed to be the largest steam locomotives running in the world today. They were built in 1955 by Beyer Peacock of Manchester for the arduous 332-mile run from Mombasa to Nairobi in Kenya. This engine, No. 5914, is named *Mount Londiani*

35

pumped back into the boiler at high temperature and thus recycled. Even greater efficiency is achieved by exhausting one safety valve into the sixteen-inch diameter pipe, so preventing loss of potentially valuable water. The design's excellence is further shown by the entire condensing cycle being completely self-sufficient and needing no attention from the engine crew; all energy being provided by the exhaust steam.

The South African Condensers are one of the last 'steam greats' and, though specialized in application, provide yet another basis of thought on what a continued steam policy might have produced. Whatever seeds of inspiration germinate, they will be sadly abandoned to academic surmise.

Condensers Today

All ninety of South Africa's Condensers remain on the active list, being found principally on main lines throughout the Orange Free State. They are now entirely absent from their traditional habitat, the barren and inhospitable terrain of the Karroo Desert, to which they owe their origin. As recently as 1974, Condensers were operating almost all the fifty-six trains daily across the desert between De Aar and Beaufort West, but in one swiftly implemented diesel take-over they disappeared within a few months.

Many wandered on to the trunk route between Kimberley and De Aar working turnabout with their non-condensing relations, whilst lesser active examples have been relegated to shunting duties, including one which is regularly assigned to shunting a goods yard adjacent to a hospital – the soft jet-like whirr of the turbine fans presumably being regarded as a greater aid to somnolence than the conventional engine's staccato bark! Their future is now questionable, the

Plate 23
A 2-truck vertical-cylinder Shay locomotive from Lima in 1912 works logs over the famous 2′ 6″ gauge Ali Shan Logging Railway in Taiwan

condensing apparatus, apart from being costly to maintain, is becoming trouble-some with age, and already several engines have been rebuilt into non-condensers at Salt River Works, Cape Town.

Crane Engines

Crane engines came to prominence in the latter half of the nineteenth century and were yet another variant which 'put asunder' the steam locomotive's traditional shape. They were principally employed in industrial environments where heavy and awkwardly-shaped loads had to be handled. The type illus-trated by plate 15 was among the last surviving examples in Britain. These little trojans worked at Doxford's Wearside shipyard carrying prefabricated parts of ships from manufacturing bays down to the fitting-out berths on the river. Originally produced in 1902, further examples were built during the Second World War, when an increase in shipbuilding necessitated more locomotives on yard service. Doxfords are famous British shipbuilders and it is interesting to note in passing that H. W. Garratt, creator of the Garratt locomotive, worked for this company during the 1880s.

Functional though the Doxford engines were, they might have been designed by Roland Emmett. Look at their massive dumb buffers on which sit a line of diminutive oil cans; a lamp little bigger than that used on a bicycle, and a long slender chimney overshadowed by the enormous lifting jib with three grappling hooks situated at specific lengths.

Although now retired from Doxfords, several examples have been preserved. Today crane engines have largely disappeared, having been superseded by such things as mobile cranes and fork-lift trucks, but odd examples do still survive in commercial service.

Plate 24
A South African Condenser 4–8–4 in the Karroo Desert. Notice the enormous condensing tender some 60 ft long and also the 16 in. diameter pipe placed along the side of the engine to carry the exhaust steam to the condensing elements in the tender

Plate 25
Following pages: Insular Lumber Company No. 7, a 3′ 6″ gauge Baldwin 0–6–6–0 Mallet of 1925 draws teak from the mountains of Negros Island. This enormous compound Mallet is undoubtedly one of the most fascinating steam survivors left in the world today

3 Modern Power

Modern Engines

Certain areas of the world will perpetuate steam power for many years to come, but inevitably the continual dwindling of variety must produce a situation in which the majority of survivors will be recently built mixed traffic engines: these, in essence, can look very similar the world over.

Until recently, the accent of railway enthusiasm was upon a wealth of diversity within the individual's own country. It was never fashionable to look at the world's railways as a whole, nor indeed was it particularly necessary, since the interest to be gleaned from one's own country was more than enough to stimulate and satisfy. Providence decreed that this insular, yet rather perfect state of affairs had to end and with the rapid demise, if not the complete annihilation of steam – as in Great Britain, America and Scandinavia for instance – the enthusiast either had to bury himself in history books, undertake local preserva-

Plate 26
Hsinchu depot on Taiwan's west coast main line acts as host to three classes. Left to right: Taiwan Government Railways Class DT 580 2–8–0, Class CT 150 2–6–0 and Class DT 650 2–8–2. These are the Japanese National Railway classes 9600, 8620 and D51 respectively

tion work, or look ahead to new horizons in an effort to regain the pleasure he had lost.

Now, for the first time, locomotives are being widely regarded on a global scale, the diversity has returned. Concurrent with this shift in our subject's status, the world has suddenly become a smaller place; not least by means of the remarkably frequent and fast global air routes which have come to such prevalence over the last two decades. It is now possible, even within the scope of a three-week vacation, to see and enjoy locomotives from distant lands which, only ten years ago, were to the vast majority of us both unknown and unimagined. Even the more sedentary enthusiast is able to take a delight on a wider basis; after all, when steam was in its heyday, few of us ever ventured to all corners of our own countries, but we were interested in the engines and gained pleasure from the knowledge that they were there – whether or not we might ever see them. How many Englishmen ever witnessed a Great North of Scotland D40 class 4–4–0 in action? But they knew about them and indeed credited them as being perhaps the best-looking engines in the land!

Thus by broadening our horizons, we reconstitute the loss of variety within our homelands wherever they be. This has been the major saving grace in the steam locomotive's decline, for had it not occurred we would almost inevitably lack the very real appreciation of world locomotives which is beginning to emerge today.

Many of us have had consciously to learn to enjoy foreign engines – especially English people weaned and bred on the smooth, handsome and orderly products of their homeland. English people are still saying: 'But foreign engines are so

Plate 27
A thoroughbred in modern express passenger locomotives – this handsome Japanese Pacific working today in Taiwan

Plate 28
Following pages, left: Standard modern power for South Africa's secondary routes are these British-constructed 'Berkshire' 2–8–4s. One is seen here heading a Port Elizabeth–Uitenhage passenger train

Plate 29
Following pages, right: A pair of South African Railway Class 15AR 4–8–2s head a goods train into Port Elizabeth. These engines, depicted in re-boiled form, were originally exported from Britain in 1914

43

Plate 30
Previous pages, top left:
In the abandoned steam
depot at Patricroft in
Manchester, lines of ex-
L.M.S. Stanier Class 8F
2–8–0s lie derelict prior to
being towed away to
breakers' yards

Plate 31
Bottom left: A more
invigorating scene at
Witbank depot in the
South African Transvaal
where a Manchester-built
4–8–2 of 1920 is seen
confronting two
transatlantic 4–8–2s of the
mid-1920s. Classes 15AR
and 15CA respectively

Plate 32
Right: Amid the smoky
intrigue of Nogent
Vincennes depot in Paris
can be seen the last of the
ex-Est Railway Class
141TB 2–8–2Ts

ugly, I could never enjoy them in the same way: our engines are so handsome –
so civilized!' But let it be remembered that attraction is largely based upon
familiarity and this must be given time to evolve. Only the exposure is needed.

Twenty years ago few people were more inward-looking than myself. I could
not have identified one 'foreign' class of engine while, even worse, I would not
have wanted to! Eventually some well-directed coercion from friends led me
over the water to France, but I was awe-struck by the creatures I found rather
than endeared. Even the now superbly handsome French Pacifics were remote
beings of another world, and despite a dynamic journey behind one down the
ex-Nord main line from Calais to Amiens I was, within several hours, homesick
for a Stanier 'Black 5'. But fortunately a little of the magic stuck; my curiosity
was irretrievably aroused, and now I can truthfully say that I am able to
appreciate world locomotives on an equal basis. Obviously there are favourites,
but drawn from the widest possible canvas – any of Karl Golsdorf's ex-Austrian
Empire engines set my pulses racing as fast as any of the old, indigenous
home products.

Possibly British and American enthusiasts find the assimilation of world
engines easier because of each country's distinctive school and vast export
market. Perhaps the same might be said of Germany. It is certainly with great
excitement that I discover the British pre-grouping look, be it Victorian,
Edwardian or even later, which so magnificently soldiers on in certain countries
today.

But what of the future? Those last ten years when interest in world steam will
probably be greater than ever before. On the main lines, traffic is likely to be

dominated by the modern utilitarian mixed traffic engine, which so often lacks not just a family delineation but also the flamboyant allure of many earlier types. Obviously our mixed traffic engines represent a fusion of design practices with the good points amplified and some inherent defects bred out. Unfortunately they can tend to look monotonously the same the world over. Understandably, recent economic problems facing the world's railways necessitated a continuous utilization of motive power and the production of mixed traffic designs capable of successfully operating a wide range of traffic. Much latter-day design was orientated towards this end. When the old lineages disappear and these faceless utilitarians take over what is left, then the funeral dirge for steam might be said to have truly begun. At present this is not so, for the modern engine simply adds another dimension to an already wide tapestry.

Our utilitarians will probably be the 2–8–2, 4–8–2 or 2–10–0 with a mean driving wheel diameter of 5′ 3″: powerful engines, with adequate adhesion, liberal axle loading and capable of some reasonably high speeds. Two outside cylinders will be only one example of their ultimate proportions and fine balance of working parts. Perhaps all in all, they are rather more Americanized than many would like to admit.

We conjecture that these 'mixed traffics' which have invaded the surviving steam ranks with a Viking-like onslaught might soon dominate the whole. Perhaps our last main line journeys will be behind a 4–8–2 whose crisp no-nonsense two-cylinder exhaust beat will reverberate with precision as she spins her 5′ 3″ driving wheels over a landscape dominated by the technology of a

Plate 33
Opposite: The modern standard mixed-traffic engine for South Africa is depicted by these splendid Class 23 4–8–2s seen heading a 2,000-ton train along one of the busiest steam operated lines in the world, that between Kroonstad and Bloemfontein in the Orange Free State. The 23s, along with the closely related Class 15F 4–8–2s, are the most numerous class ever to run on the African continent; there are, in total, almost 400 engines, all of either British or German origin

Plate 34
Above: The ultimate in Spanish steam design came with the ten 'Confederation' 4–8–4s built by Maquinista of Barcelona in 1956. After taking water, one proceeds to the adjacent goods yard to collect its train

49

totally different era. And into the twilight she will go, and against the ever darkening sunset will be seen her familiar regularity of shape: low boiler mountings, large windshields, and a wide gap between her long slender boiler and frames. There might evolution's imagination finally draw the epoch to its ultimate fulfilment.

Modern Engines Today

In some countries the remaining steam turns are almost exclusively handled by the kind of locomotive previously described: Spain is a typical example. The antiquities possessed by Spain up to ten years ago have already been mentioned, but now such steam traffic as remains is powered by the 141F class 2–8–2s– two hundred and forty having been built between 1953 and 1960. Some parallel can be found in the 1,343 American-built 141R 2–8–2s which are just bringing the steam age to a close in France. Superficially resembling the 141Fs, these were even more unsophisticated, especially in appearance, and although they performed superbly they brought a sad anti-climax to French steam – a school always noted for elegance and a profundity of mechanical sophistication. Portugal however, provides a marvellous contrast: no such utilitarian engines exist, and at least a dozen classes remain active with little more recent than 1920 on any of their worksplates.

Most European steam today is found in the communist bloc states, especially Poland, East Germany and Czechoslovakia, though Turkey also retains a dense and highly active steam fleet. The 2–8–2, 4–8–2 and 2–10–0 abound throughout these countries and provide a backbone of motive power. The 2–10–0 is especially represented by innumerable ex-German war engines – *Kriegsloks* which, subject to many modifications of detail, are still well in evidence – their all-round usefulness in wartime simply being echoed by the requirements of modern times. Over six thousand were built, some actually after the war – plate 81. The only steam engines truly indigenous to Yugoslavia were the one hundred German-built standards composed of Pacifics, 2–8–2s and 2–10–0s. Today these classes, along with a proliferation of German war engines, constitute the bulk of this country's steam roster.

Finland keeps a small but well maintained steam fleet now principally composed of the sixty-seven handsome Tr1 class 2–8–2s which were built up until 1957. Neighbouring Russia has for long since had a ruthless standardization policy, well evidenced by their SO class 2–10–0 – some five thousand were built.

India is still heavily reliant upon steam power and she, after a gloriously diversified past, has now settled for two brilliant standard designs: the WG class 2–8–2 which now number two thousand units, and the closely related WP Pacifics which number in excess of five hundred engines. The WGs have the distinction of being amongst the very last steam engines to be built, with construction continuing into the 1970s. A situation of comparable latter-day standardized building exists in China and it is widely believed that she, alone in the world, perpetuates steam construction today. It is difficult to make generalizations about China, but steam usage is known to be widespread. Even the working museum in Indonesia possesses one hundred Krupp-built 2–8–2s of 1951. Typical in appearance, these are by far the country's most numerous steam class.

South America's vastness, with all its rich diversity, can still offer a tremendous

Plate 35
Left: A South African Condensing 4–8–4 dashes southwards through the Karroo Desert with a coal train bound for the Cape Province. In the background can be seen the Three Sisters, a famous Karroo landmark

Plate 36
Below: An East African Railways 'Governor' Class 4–8–2 + 2–8–4 Garratt ambles along the single track line connecting Tanzania with Kenya with the overnight freight. Governors were built in the early 1950s; they are modern lightweight power for East Africa, with a remarkably light axle loading of only 11 tons

Plate 37
Following pages: Near the Russian border in north-eastern Finland are found the last survivors of the Finnish Class Tv1 'Jumbo' 2–8–0 heavy goods engines. Built with an axle loading of 13·1 tons, these engines have been very important freight power in the country for some 50 years

range of steam locomotives and, taken in its entirety, is the most prolifically varied region in the world for the steam enthusiast. Old and new work side by side, though surviving pockets of steam are often extremely widespread and difficult to reach. Few classes on the whole South American continent number more than a handful of engines, though modern mixed traffic types, particularly of American and German origin, are widespread.

South Africa has become justifiably world famous for its great steam main lines where large engines are put through their paces with heavy trains. The country still sports a wide range of classes, though nowadays work is becoming increasingly entrusted to the 1930s-built 15F/23 class 4–8–2s. Together these similar engines total almost four hundred units and, taken collectively, are the most numerous class ever to run on the African continent. However, some African countries are notable exceptions to the rule: Rhodesia's Garratt fleet has already been discussed, and there the last steam train will doubtless draw to a stand behind a Garratt. Moçambique, Angola and East Africa all have many Garratts in action, though in the last mentioned modern mixed traffic Tribal classes of both 2–8–2 and 2–8–4 wheel arrangements will doubtless be the last steam to remain active; over one hundred of these fine engines were built between 1951 and 1956.

In general then, it is the modern mixed traffic engine which will weave an increasingly discernible thread of solidarity and uniformity throughout the surviving ranks of world steam; against this thread the old family lineages still extant will provide a blissful contrast.

Plate 38
Opposite: The 4–8–0 type has been widely used in Spain; here a very capable looking Class 240F brews up at Salamanca. She is one of the earlier ex-MZA engines built from 1920 onwards. Altogether some 300 engines made to this basic design were put into traffic and they formed an important backbone to latter-day Spanish steam traction

Plate 39
Above: This ultra-modern East African 2–6–2T built by Bagnalls of Stafford in 1951 is seen at Tabora in Tanzania. She is one of two such engines originally intended for dock shunting at Dar-es-Salaam. The Giesl chimney is a recent refinement

55

4 Engines of Heavy Industry, Sugar Plantations and Lumber Companies

Industrial Engines

The steam locomotive was born into industry in 1804 and in terms of commercial operation will probably die there. Its flowering on to a main line occurred some twenty years after its inception, and likewise, at the opposite end of the epoch, its final moments may well be enacted in an industrial setting many years after the main lines have been modernized.

Understandably, the last few years have witnessed a plethora of railway literature greatly motivated by the decline of steam. In fact, the impetus thus generated seems to have gate-crashed into the diesel age as well, with impassioned articles, and even books, devoted to the final chunters of various classes a mere twenty years old! Yet, in our more pensive moments, we might wonder how close the disappearance of steam actually is – particularly when considering the complex questions involved in providing a fast developing world with the energy it demands. Anyone with a quick and easy answer to such questions is more likely to be a fool than a genius, but whichever way fate pushes the tech-

nological pendulum, the industrial steam locomotive at least has a distinct future – certainly long enough for the millions of people who revere and gain pleasure from an operating steam complex to set by sufficient time and money. There are times when our debilitating obsession with the end of steam gives false bearings, and although it is customary for educational establishments, and indeed many industrial institutions, to regard steam as a fragment of deep history, it is still one of the world's great prime movers, and we who possess regard for the subject are all the better for knowing it. Only in 1971 did the Hunslet Works of Leeds build a new steam engine for the Javan Sugar Plantations, while in 1974 an industrial unit in Sumatra was anxiously scanning the world for five new steam locomotives. This company had an abundance of wood in the area; they wanted steam engines and nothing else. Isolated examples concerning remote backwaters of the world these may be, but so prolific is industrial steam today that we can cool the fever and relax a little – some of it will see the century out.

Industrial engines have attracted many devotees, not to mention the few dedicated specialists whose knowledge and interest in railways spans little further. Certainly they belong in a different perspective or 'order' as a naturalist would say, since apart from being built almost entirely to the specifications of industry they are invariably smaller. Due to a wide range of builders – over two hundred have existed in Britain alone – the variations in design have been legion. In common with main line engines, this diversification of design and form culminated in similar power outputs – an ultimate paradox, which, though frustrating to the rationalist, has provided that epicurean delight, variety. But

Plate 41
Left: The great Corby Steel Works set in the heart of the Northamptonshire ironstone bed, seen here on a cloudy afternoon from an adjacent ironstone mine. Industrial settings such as this still act as host to steam engines in many parts of the world

Plate 42
Above: The evocative background of a slag tip provides a typical foil to a Hunslet Austerity 0–6–0ST as it undergoes its workaday chores at Maesteg, a surviving outpost of steam traction amid the collieries of South Wales

59

unlike many main line designs, there have been few absolute standards in the industrial world, even among the big manufacturing companies, for although many had their own basic designs and shapes, they often managed to make the most delightful twists and variations on the theme faithfully tailored to the needs or fantasies of the particular recipient – albeit that he might be buying only half a dozen engines or even less. Even when standard designs were distributed, the various lettering styles, liveries and other sundry aspects applied by their operators meant that very few ever looked alike.

Cross-fertilization of the two orders can occur in several ways, though particularly when engines become either obsolete or superfluous on main lines, resulting in their sale to industry for a further lease of active life. Such engines are often immediately recognizable, though not always. An example can be given concerning *Eslava*, a 5′ 6″ gauge 0–6–0 built by Hartmann of Chemnitz, Germany, in 1881 for passenger work on the Asturias – Galicia – León Railway in Spain. Some years later this railway was absorbed into the NORTE – one of Spain's principal main line companies, and thus *Eslava* worked for over fifty years until, in 1951, the country's railways were nationalized into the RENFE or Spanish National Railways. So the little veteran worked for more years, until modernization led to her being sold to a private colliery company in the north-western part of Spain. *Eslava*, now exclusively a coal engine, soldiered on through various inter-colliery transfers until in the mid-1970s, nearly one hundred years later, she was discovered working at Ujo Colliery and still capable of buffeting an appreciable rake of coal waggons around the mine.

Natural adaptation is another way in which this cross-fertilization can occur,

Plate 43
Left: A delicious Orenstein & Koppel 1914-built outside-framed 0–8–0TT of 700 mm gauge, shoots fire as she passes a yellow flowering tree. The engine, which burns bagasse, works for the Purwodadi Sugar Factory in Java and is named *Bromo* after an extinct Javan volcano

Plate 44
Above: The last surviving 'Workhorse' 0–6–0ST, built by the Yorkshire Engine Company of Sheffield, hard at work at the Clipstone Colliery, Notts

a good example being the North British-built South African Railway class 12A 4–8–2 main line coal engines introduced in 1919. After the Second World War these were perpetuated, in unsuperheated but otherwise identical form, by the North British for the private colliery companies of the Transvaal. Similarly, the successful South African Railway 19D class 4–8–2s were adopted by the Rhodesia Railways, and when the R.R. sent their order to Henschel in Germany they were requested by the Wankie Colliery Company to add four extra engines on, in unsuperheated form, for use around the colliery network. Many similar instances occur, though usually main line and industrial engines are designed and built quite separately.

The various schools of practice discussed later in this volume can be applied in very broad principle to the industrial engine also. Some emergent forms do appear despite the opaqueness of the overall picture. The three major producing countries have been Great Britain, Germany and America: each to some extent has perpetuated their national appearance, though designs originating in one country may, years later, be perpetuated in another, when the industrial user requires more engines. But whatever background influences have filtered down into the industrial engine, they only serve to accentuate its individualistic nature and this, combined with both gauge and environment, renders it all the more worthy of serious appreciation and study.

Possibly the best justice that can be done to this subject is to discuss some engines along with the systems and environments in which they are found. Only a small selection can be included here, but the ones I have selected are typically representative. Therefore, if we work our way through the pictures contained in this chapter, we might try to evoke some of the industrial and rustic atmosphere so poignantly conjured up by these systems. As Charles Small has pointed out, the industrial steam engine, though needing to be understood within itself, should never be seen in isolation from its working environment.

Corby

One of the finest industrial steam complexes in Britain was the standard gauge ironstone network based on Corby Steelworks. This system was like a miniature railway company. From the Steelworks, set in the heart of the ironstone bed, lines radiated outwards in all directions like spokes from a wheel hub, each terminating in an ironstone mine where the tracks would be slewed up against a huge digger. All the mines were opencast, though it was not unusual for ore to be extracted from deep gullets of up to seventy feet, and even more. Operations around the Steelworks itself were performed by Hawthorn Leslie 0–6–0STs, painted bright yellow to make them more visible amid the gloomy works surrounds. Out on the mines 0–6–0STs were also used, but varied from diminutive Manning Wardle veterans of 1895 to large eighteen-inch cylinder engines specially built by Robert Stephenson & Hawthorn for the mineral lines up to 1958. All were painted green.

The mineral lines' rustic atmosphere made them doubly exciting. They wandered far out into the Northamptonshire countryside past small villages built in sandstone; wild life abounded, with blackberries and wild strawberries growing in abundance on the tangled embankments. The uneven trackbeds twisted through meadow and coppice alike, some lines only carrying two or three trains a day. Imagine those little green engines barking their way over the countryside hauling long rakes of waggons piled high with red ironstone as, passing through an ever changing backcloth of woods, fields and villages, they tinged the pure country air with acrid whisps of locomotive smoke.

Against the mines' rustic charm, contrast the intense clamour of the great Steeelworks. Here amid a never ceasing roar of activity and bustle the yellow Hawthorn Leslies could be seen trundling waggons laden with molten iron against the furnaces' heavy incandescent glow: this was the nerve centre, whose one aim was to produce a million tons of steel a year.

Sadly, this particular system has now succumbed to dieselization, but many similar concerns do survive with steam. Plate 41 depicts some of Corby's intense atmosphere.

The Javan Sugarfields

I suppose the Javan Sugar Plantations must rate as one of the finest industrial complexes left anywhere today. On Java there are fifty-two sugar factories, the majority set among big plantations, and the railway operation is similar in principle to the one described at Corby. Although most factory networks are not interconnected they are, from an enthusiast's viewpoint, tantamount to being one big system, since most are situated in central and eastern Java and are thus densely spread over a relatively small region. The majority are state controlled. Since Indonesia's national railway is almost a working steam museum, one might imagine the most veritable gems to be tucked away amid the plantations but, disappointingly, this is not the case as most sugar factories have been developed within the last sixty years. Doubtless many State Railway antiquities would have passed to sugar lines had the gauge been compatible, but most Javan sugar networks are laid to seven hundred millimetres against the 3′ 6″ of the main line. Nevertheless, for constant action, variety of power and an environmental beauty, studded with tropical vegetation and extinct volcanoes, a visit to these plantations is a must.

Those who can happily lose themselves amid these networks will discover many engines, the majority of German origin. Although the motive power is highly varied both in livery and design, the 0–8–0 tank engine with a large bagasse-holding tender attached predominates. *Bromo* on plate 43 is a splendid example gloriously decked in yellow and red. In accordance with Javan practice, she is named after an extinct volcano. Many other types can be found, including some plantation Mallets.

The typical engine roster of an average-sized factory is epitomized by P. G. Lestari near Kertosono. This factory operates eight locomotives of 700 mm gauge, set out as follows:

Engine Number	Wheel Arrangement	Builder	Works Number	Date
1	0–6–0TT	Orenstein & Koppel, Germany	(Unknown)	1917
2	0–8–0TT	Orenstein & Koppel, Germany	6873	1913
3	0–8–0TT	Orenstein & Koppel, Germany	10443	1923

Plate 45
Following pages: A splendid chocolate-coloured 4–8–2T, built in 1951 by the North British Co. of Glasgow blends harmoniously with the industrial landscape of South Africa's Transvaal. Quite a number of these handsome engines are still at work in the South African collieries

4	2–6–2T	Orenstein & Koppel, Germany	11638	1928
5	0–8–0TT	Orenstein & Koppel, Germany	9360	1920
6	0–4–4–0TT Mallet	Borsig, Germany	10936	1921
7	0–4–4–0TT Mallet	Orenstein & Koppel, Germany	11542	1927
8	0–6–0TT	Orenstein & Koppel, Germany	9221	1920

(TT = tender tank engine, the extra fuel space being necessary to accommodate the large quantity of bagasse needed)

At least three hundred and fifty steam engines can be found active during the sugar season, which reaches its peak around June.

The Philippine Sugarfields

Sugar provides the principal dollar-earner for the Philippine Islands, and one of their finest concentrations of sugar growing is on Negros Island. Negros is much smaller than Java and only supports ten companies, but each has a large factory set amid a vast plantation. It is interesting to compare the overall difference between the locomotives of these two islands: Java has complete orientation towards Europe, whilst Negros is almost entirely American in practice with the stalwart 2–6–0 tender engine from Baldwin/Alco predominating. When Javan sugar industries were being developed, the island was part of the Dutch East Indies: in contrast the Philippines were developed under American control. One exception occurs on Negros with the Victorias Milling Company, whose roster is partly German – the result of an intensive German campaign to expand their far eastern locomotive exports. Nearly all factories on Negros are laid to a three foot gauge and mill between October and April, though Victorias Milling, which operates over three hundred and thirty kilometres of 2′ 0″ gauge track, works all the year round.

A really exciting locomotive roster on Negros is that owned by the Hawaiian-Philippine Company – a 3′ 0″ gauge concern possessing some one hundred and thirty kilometres of track. Their engines, known as Dragons, are in permanent radio-control with the factory. See both the table below and plates 56, 82, 83.

Dragon Number	Wheel Arrangement	Builder	Works Number	Date
1	0–6–0	Henschel, Germany	19688	1923
2	0–6–0	Baldwins, America	52199	1919
3	0–6–0	,, ,,	52864	1920
4	0–6–0	,, ,,	52865	1920
5	0–6–0	,, ,,	52866	1920
6	0–6–0	,, ,,	52867	1920
7	0–6–0	,, ,,	60677	1928
8	0–6–2ST	,, ,,	(Unknown)	1924
9	0–6–2ST	,, ,,	(Unknown)	1916

Plate 46
Opposite, top: Dawn at Backworth, north of the River Tyne, finds a Hunslet Austerity 0–6–0ST busily hauling a rake of coals away from Eccles Colliery

Plate 47
Opposite, bottom: This period piece is an inside-cylinder 0–6–0 side tank built by the Hudswell Clarke Foundry of Leeds in 1909. She is seen at the Bedlay Colliery, Glenboig, near Glasgow

Plate 48
Above: Bedford power
station provides an exciting
backdrop to an oil-fired
Andrew Barclay 0–4–0ST as
it backs a loaded coal train
into the power station.
Although the engine was
built as recently as 1953, the
design is to a basic pattern
prevalent in British industry
throughout the 20th century

Dragons 3 to 7 are simply enlargements of No. 2 and have twelve-inch cylinders against the earlier engine's ten inch ones. Dragon 1 is a Europeanized hybrid based on the Baldwin engines of 1920, though when No. 7 was built in 1928, the Company reverted to Baldwins.

Dragons 8/9 are small saddle tanks which were transferred from the Company's Hawaii plantations after dieselization there. These engines, along with No. 2, chiefly work around the factory, leaving the larger engines to operate the plantation lines.

I shall try to give some idea of the transport and loading method. Water Buffalo are employed to bring the cane up to the rail connections set throughout the plantations. The cane is then manually transferred on to the rail waggons. These Water Buffalo work superbly on cane haulage, but periodically they like to rest for an hour or two completely submerged in water with just their noses and eyes protruding above the surface. Afterwards they are content to haul cane for another shift lasting several hours. Possibly once every twenty-four hours, the engine will draw loaded waggons out from these connections and push a replacement rake of empties in. Some of my finest railway adventures came when I rode out with these trains at all hours of the clock, operations being highly intensive throughout the milling season.

On one such run we were scheduled away from Ma-Ao Sugar Factory at 02.15 hours and, arriving at the yard in good time, I found old Alco 2–6–0 No. 5 being loaded with bagasse. We were booked along one of the mountain routes which was to take us to the foot of an active volcano: because of their superior braking capacity the steam engines are used over the mountain lines in preference to diesels. Shortly after 02.30, No. 5 drew up behind a long rake of empties and, after a short bout of slipping, she propelled them vigorously away from the yard. Within minutes we had left the dimly lit factory behind, and passed into the dark remoteness of the plantation. It was one of the clearest nights I had ever seen. The volcano, although some fifteen miles away, loomed up with remarkable clarity against a moonlit sky literally covered with stars.

The climb began almost immediately. No. 5 bit into the grade and our two firemen frenziedly fed bagasse into the firebox. Bagasse is a natural waste product of the sugar cane processing, for when the cane is crushed and the juices are extracted the resultant fibres are dried out and baled up into large cubes for use as locomotive fuel. Thus, in effect, the engines are operated at remarkably low cost. Our engine's valves were set to 160 lbs per sq. inch and the needle hovered precariously below that figure. The calorific value of bagasse is so low that even with two firemen it is a struggle to maintain a full head of steam – especially when climbing hard. Of course, in countries where labour is plentiful it is cheaper and better to burn bagasse than to purchase fuel, be it wood, coal or oil.

The plantation itself was in complete darkness, and some miles ahead it was possible to discern the russet fireglow and blazing headlamp of engine No. 2 – another Alco 2–6–0 which had left the factory an hour before us and was sharing our track until a divergence point actually at the volcano's base. The coolness of the night outside did little to assuage our firemen who, with sweat glistening on their brows, fought against tremendous odds to maintain boiler pressure. The enormous lumps of bagasse being flung into the firebox exploded almost instantaneously, like a match to dry straw, and seconds later a defiant flurry of

Plate 49
Below: In marked contrast and on the other side of the world, an old Lima 3′ 6″ gauge 3-truck 3-cylinder Shay draws a long line of teak logs over the metals of the Insular Lumber Company on Negros Island, Philippines

Plate 50
Cadley Hill Colliery in the English Midlands is a noted outpost of industrial steam traction. Featured is *Cadley Hill No. 1*, a Hunslet Austerity 0–6–0ST of 1962 and one of the last three ever built, whilst on the left is *Progress*, built in 1946 by R.S.H., and the last survivor of an old Hawthorn Leslie design dating back some 60 years. On the skyline is Drakelow power station, one of the largest in England

sparks shot skywards from the balloon-stacked chimney. She was eating it without bothering to swallow it, and pressure was down to 130 lbs per sq. inch.

Suddenly a swinging red light flashed from the leading waggons some fifty yards ahead: the brakeman's signal for an emergency. Our driver slammed the regulator closed and made a full brake application, causing both myself and one of the firemen to lose our balance. The first five waggons were off the road and two of them had ended up at right angles to the track and buried themselves amongst the sugar cane. Imagining that this would cause several hours' delay, I was greatly surprised when, in a mere fifteen minutes, all offending vehicles were re-railed with the assistance of some expertly manipulated re-railing clamps. It was explained to me that derailments occur on every trip and, while they can occasionally take hours to correct, the brakemen have become so adept at handling them that trains are usually on their way again within twenty minutes. Plate 89 shows a scene from this journey.

With No. 5 actually blowing off, we continued our journey, and as we pounded over the last few miles the first flush of sunrise appeared behind the volcano – by now immediately ahead of us. The ginger sunlight caused our side of the mountain to change from black into a successive progression of purples. When the sun finally burst over the edge it lit up the grey crater and lava-ridden wastes into their true perspective.

After an enthralling four hours collecting cane from remote sidings, we headed our loaded train towards the factory, and it was on this return journey that a really terrifying event occurred. Perhaps the magnitude of my fear was conditioned by the many disaster tales the men had recounted during our

70

outward trip for, despite the obvious aesthetic pleasure, railroading in these remote areas is a hard and dangerous business.

I was unconcernedly riding on the back of the tender eating sugar cane – the deliciousness of which has to be experienced to be believed – when, looking back along the train, I realized to my horror that we had become separated from the rake, and that the entire train was bearing down upon us some twenty yards back and, by virtue of a long down gradient, was approaching at a much faster speed than we were moving. I had heard of this happening, sometimes with the result that, upon striking, both train and engine are derailed. Even worse, if speed is high enough, the runaways can rear up and crash down over the engine. Realizing that the engine crew had not noticed the breakaway, I flung myself across the tender top and called down to the driver. Apart from No. 5 rattling badly she was also blowing off, and he could not hear. Not daring to look behind, I yelled again. Instinctively, the driver realized what was happening and he accelerated the engine with all the drama of a scene from *The Great Locomotive Chase*. Well, that runaway rake of sugar did catch us, but at an almost equal speed and all we felt was the gentle bump of couplings before No. 5 drew her train to a standstill. The crew looked pale and I was actually shaking – so close had we been to an accident.

Within seconds we were all laughing and one of the shunters, who spoke little English, gave us a hilarious mime indicating how the waggons might have risen in the air and scattered in all directions. We examined the leading waggon and found that the steel coupling attaching it to the engine had snapped in half as if it were a matchstick. Presumably, excess buffeting over uneven trackbeds had caused the coupling to weaken and break. The track's undulating nature had enabled the engine temporarily to pull away from the train on a short up-grade but, once the gradient turned downwards, the waggons gained a terrific momentum – this in turn being made worse since No. 5 had slowed down, as a particularly rough stretch of track lay ahead. Replacing the coupling from a reserve supply kept on the engine, we continued our journey. Plate 91 was taken as we neared the factory half an hour after the breakaway occurred.

The Workhorse

A romantic testimonial to the last surviving British 'Workhorse' 0–6–0ST is depicted on plate 44. She was one of a breed of British industrials associated with the Yorkshire Engine Company of Sheffield and, despite her bearing a 1952 worksplate, she was actually despatched from their works on 6 May 1953 to the Appleby Frodingham Steel Co, Scunthorpe – the company for whom the design was originally prepared. She worked at Scunthorpe until dieselization led to her sale, along with some sister engines, to the National Coal Board.

The plate, made in April 1972, shows the engine at Clipstone Colliery, Nottinghamshire, on its very last journey. Ten minutes after the picture was made the fires were dropped and another notable British class passed to extinction.

Although essentially a British industrial design, one of the breed, identical in proportions, was shipped to Peru, where she survives today in oil-burning form on the Ferrocarril Central del Peru. This engine, exported in 1952, shunts in the port at Lima on the Pacific coast.

Coalfields and Goldfields

Another very rewarding area for industrial steam locomotives is the Transvaal of South Africa, especially the great Witbank coalfield. Centred upon the industrial town of Witbank, this region has a unique atmosphere, being unobtrusively set amid a vast tableland of grassveld and maize-producing farmland. During wintertime the green vegetation becomes bleached to a yellowish-gold, due to prolonged sunlight and absence of rain. This is the major coal-producing area for the Republic of South Africa and is, to this day, haunted by nineteenth-century British-built steam locomotives.

In contrast with many countries, South Africa's industrial engines are of 'main line' proportions, indeed many are ex-main liners, including some Beyer Peacock 2–8–2+2–8–2 Garratts. Hauls from the outlying mines up to the State Railways connection are often lengthy and arduous. It has already been mentioned that the old main line 12A class was specially built for collieries as well – these engines have twenty-four inch diameter cylinders and a tractive effort of over 47,000 lbs. Because South African collieries are not nationalized, each company adopts whatever livery it will, and it is possible to see the 12As decked out in black with yellow lining, maroon, Caledonian blue and green.

Plate 54 shows a 12A in typical Witbank settings; notice the uninterrupted landscape stretching away to a distant infinity and the enormous open sky – an environment radically different from that of many colliery networks. The engine is seen biting into a 1 in 100 bank with a twelve-hundred-ton coal train: such duties necessitate the use of powerful locomotives. A maroon 12A is seen on plate 69. Notice how she is blowing off simultaneously with the fire being coaled up. This is not inefficient enginemanship because the locomotive is anticipating a heavy climb one mile ahead, for which combustion must be complete and maximum heat available for the long slog. If she fails to make that grade, a three mile reverse into the colliery yard has to be undertaken and the entire effort begun again.

It is heartening that so modern a technological country as South Africa can still produce a variety of colourful and multi-origined steam engines. Even today, the industrial lines see little in the way of standardized classes, though one type, which can be regarded as an industrial standard, is the extremely handsome North British 4–8–2T depicted on plate 73. Many of these were delivered to collieries over the late 1940s/early 1950s. Another example is seen on plate 45, this time in a splendid chocolate livery, and she, in common with the industrial backdrop, blends well with the golden vegetation.

As if the variety of engines was not enough, the South Africans have a habit of 'hybridizing' their tank engines – a practice which began long ago on the main lines and one which many industrial concerns have copied. This 'hybridizing' takes many forms, but frequently involves making tender engines out of tank engines, and quite a number of the North British 4–8–2Ts, despite their modernity, have suffered this fate. The side tanks, rear trailing axle and coal bunker are all removed and an old tender, usually nineteenth-century in origin, is added in place. To reconstitute the adhesion loss caused by removing the side tanks, false splashers containing three to four tons of scrap metal are fixed to the running plate on either side of the engine. Thus a handsome and modern-looking 4–8–2T in the best of British traditions becomes a rather nondescript 4–8–0 tender engine looking vaguely like a British colonial export from the

Plate 51
Opposite: This standard 18 in. cylinder 0–6–0 Side Tank, built in 1913 by Andrew Barclay of Kilmarnock, works over the historic Waterside colliery system in Ayrshire. Originally built for the Dalmellington Iron Company, she is still working over the same metals more than 60 years later, now under the auspices of the National Coal Board

1920s! If, however, a home-made tender is added instead of a proper one, the unfortunate locomotive ends up looking more like a freak. It seems that different companies hybridize for different reasons. Some simply wish to increase coal and water capacities for long runs, while others claim that leaking side tanks promote excessive slipping. One engineer maintained that tank engines slip excessively through lack of adhesion when the water level is low, especially when hauling heavy trains, though another company claimed that water surging on either side of the engine caused excessive axle box wear. Doubtless many of these reasons are valid – each company having its own particular needs and operating conditions. But whether modern-day power, hybrid, or genuine veteran, the industrial lines of both the Witbank and Natal coalfields offer a splendid range of locomotives.

The world-famous three-hundred-and-thirty-mile 'Golden Arc', which contains most of South Africa's goldfields, is situated principally in the Transvaal. Here again, many fine engines survive, often of old main line status. Plate 52 shows an ex-Natal Government Railway 'A' class 4-8-2T working for the Grootvlei Proprietary Mines in Springs. One hundred of these engines were delivered by Dubs of Glasgow between 1888 and 1900. Three of these veterans now work this gold mine, having passed from the main lines during the 1930s, and the Company claims that, between them, the engines have already moved some forty-one million tons of gold-bearing ore. Such can be the utility of old main liners sold to industry.

Without doubt the coal and gold fields of South Africa's Transvaal will remain a great attraction for some years ahead: the many classes and livery schemes being further invigorated by bold and colourful lettering gaily applied by the numerous companies.

Hunslet Austerities

The ubiquitous Hunslet Austerity has done much to perpetuate the life of industrial steam in Britain, and many collieries which would have long since turned to diesels have maintained a standardized steam roster of these sturdy little 0–6–0STs. For years it was fashionable to regard them as just another wartime design, devoid of any distinction other than being strictly utilitarian, but nowadays almost everyone accepts their radiant personality, for in their own right, they are very pleasing engines to behold. See plates 42, 46, 50.

Their advent occurred during the last war when the Ministry of Supply was seeking a light all-purpose engine to follow the Allied Armies after their secretly planned invasion of France. Initially it was felt that the L.M.S. 'Jinty' 0–6–0T would provide a suitable basis, but Hunslet's chairman, Edgar Alcock, convinced the Ministry that his Company's design of eighteen-inch cylinder saddle tanks would be superior by virtue of their simplicity in design, along with a shorter wheelbase which would facilitate a greater route availability: furthermore, he claimed, the engines would be well capable of lifting eleven-hundred-ton trains over level track – a stipulation laid down by the Ministry. Hunslet's won the day, their design was accepted and so was born a memorable class in world locomotive history.

The initial engine emerged from Hunslet's famous Leeds Works on 1 January 1943, and by 1945 one hundred and fifty had been built, some having been commissioned from other builders. Many went abroad and became spread over a wide military front, but when the fighting ceased most, but not all, returned to England. Eventually these became superfluous to army requirements and a number passed into industry, especially collieries.

The National Coal Board soon realized what 'gems' these engines were and how favourably they compared with the conglomeration of ancient and un-standardized engines which the Board had inherited from the private companies. It is little wonder that building was perpetuated for industrial use and by 1964, when the last ones appeared, a grand total of four hundred and eighty-four engines had been reached, their classic shape being a familiar sight the length and breadth of Great Britain. The N.C.B.'s North-Western Division did particular justice to the class by sophisticating them with Giesl chimneys, maroon livery and such marvellous names as *Hurricane*, *Warrior*, *Revenge* and *Warspite*: a fitting tribute to these majestic rearguards of British steam.

Goldington Power Station

What motivates the naming of a locomotive? Often the reasons are straight-forward enough, though sometimes a fascinating chain of events leads up to them, as was the case at Goldington Power Station, Bedfordshire, where the two little Andrew Barclay 0–4–0STs are named *Richard Trevithick* and *Matthew Murray* respectively. Trevithick pioneered the world's first steam locomotive in 1804, whilst Murray, one of Trevithick's associates, was also a locomotive engineer of considerable status and is especially remembered for equipping the Middleton Tramway in Leeds with a steam engine in 1811. One might assume that such names had simply been applied by a railway-conscious management

Plate 53
Following pages: Seen against an ominous typhoon bringing torrential rain and high winds a 3′ gauge Alco 2–6–0 of 1921 basks in the fast diminishing sunlight. The engine is working for the Ma-Ao Sugar Central Company, Philippine Islands

75

Plate 54
Above: Industrial engines
used on the colliery hauls in
South Africa are frequently
of main-line status. An
example is this unsuper-
heated 4–8–2 with 24"
cylinders. The engine, built
by the North British of
Glasgow in 1947, takes her
design from the main line
12A Class first introduced
in 1919. This picture is set
amid the flat grassland of
the Witbank Coalfield

Plate 55
Right: Compare this
diminutive British colliery
engine with its South
African counterpart. This
Robert Stephenson &
Hawthorn 0–6–0ST with
17" cylinders works coal
across the 5-mile branch
line from Whittle Colliery
in Northumberland up to
the British Rail main line.
The engine was built as
recently as 1955 though to a
much earlier design

at the power station, but in fact the station superintendent's wife was a direct descendant of Trevithick: some of the famous engineer's possessions now resting with her family. Thus it was with great significance that these modern counterparts of Trevithick's invention took their names. Perhaps this story is the reason for steam traction surviving at Goldington after it had disappeared from most other power station networks.

In contrast with almost every other British industrial, these two engines burn oil and have worked at the power station since it was built in 1954. Compared with the embryo creations of Trevithick and Murray, the thirty-ton Barclays are of a modern standard design with fourteen-inch cylinders. Over three hundred similar engines have been built with variations of detail. .

Cadley Hill Colliery

The sentiment and nostalgia evoked by the steam engine in its declining years is a phenomenon known to everyone in the western world. On more than a few occasions such feelings have influenced policy at certain main line depots and within some industrial units as well. Cadley Hill Colliery, set in the English Midlands, is one such establishment. Here, the diesel is firmly resisted and three steamers, maintained in a superb mechanical and aesthetic condition, are responsible for a highly efficient operation. Two of these engines are sole survivors of their species, while the third is a Hunslet Austerity of 1962 – one of the last three built. Plate 50 shows operations in the colliery yard with a skyline dominated by the enormous Drakelow Power Station.

Under the colliery manager's protection, and the mechanical foreman's practical enthusiasm, the three locomotives operate today as if new, and the colliery has become an immense attraction to both British and foreign enthusiasts alike; school parties studying industrial history have also made Cadley Hill a Mecca. So this magnificent pocket of steam lives on; three different engines painted green, red and blue respectively all maintained by enthusiasm and a pride in the job.

Other Industrials Today

The systems just mentioned have been selected to provide a general insight into the atmosphere and uniqueness of industrial railways. Unfortunately, it is impossible to do more than elementary justice to them in so short a space. One particularly notable concern which has been omitted here is the Insular Lumber Company, which is now little short of being a living legend: this incredible system is covered in the American section of chapter 5.

Most industrial systems connect with main lines, but not all. Countries and islands devoid of main line railways often have prolific industrial networks. A typical example would be Negros Island in the Philippine archipelago, which has some 1,500 kilometres of track owned by twelve different companies.

It is difficult to give a succinct account of world industrial steam, partly because the subject is so vast and partly because much remains to be discovered. Russia and China are almost completely unknown areas, aliens seldom being allowed anywhere near active industrial regions. These countries, however, need not worry the enthusiast, since many parts of the accessible world also harbour totally undocumented steam complexes. Most European countries can still produce a good selection: perhaps notably the narrow gauge forestry

Plate 56
Above: It is usual practice to burn off the dead cane leaves after cutting, and against such a blaze is Dragon No. 6 of the Hawaiian Philippine Co. seen rolling empties through the plantation. Note the uncut cane in the rear of the picture and also how the train is shimmering in the heat haze. She is an 0–6–0 built in 1920 by Baldwin of America

Plate 57
Opposite: Flinging teak sparks into the tropical vegetation, old Lima Shay No. 12 of the Insular Lumber Co. shuffles her way back to the loading area to collect a trainload of teak

systems of Poland and the coal-producing regions of north-western Spain. In contrast though, both Scandinavia and the Low Countries are now almost entirely steamless.

India, along with its world-famous main line roster, can still sport a tremendous variety of industrial engines with British, German and even American engines, all highly active. These work over a splendid range of habitats, such as sugar plantations, iron and steel works and the great Indian coalfield.

The various plantations of Java and Sumatra have already been covered, but lesser known is the huge network of steam-operated sugar railways on Taiwan (Free China). Concentrated on the island's western side, these railways link up with the west coast main line and are close to another of Taiwan's irresistible attractions – the Ali Shan Logging Railway. Far away across the Pacific, the last remnants of Australia's once rich locomotive heritage are now ending their days amid the country's sugar fields.

With North America and Canada completely steamless, apart from isolated industrials, we turn again to South America. Here, widely scattered over a vast continent, the ardent traveller will have much to discover and doubtless some of it will be completely undocumented, though the Brazilian sugar plantations will be high on his itinerary.

The foregoing are but fragments of the whole and, in spite of much arduous and imaginative research by many enthusiasts, the field remains largely unexplored. But surely the contents of this chapter provide ample enough proof that the industrial steam engine will, for many years to come, provide a basis for stimulating research – hopefully followed by many action-loaded adventures.

5 Schools of Practice

Austrian Empire Engines

Inspired by the warm, sunny days of spring, a corn bunting erupted a vivacious snatch of song from the lineside wires. The wind blew lightly over an open expanse of countryside already decked with vividly coloured spring flowers, and although the distant mountains were still snow capped, the Greek spring-time had clearly begun. An overnight goods from Athens to Thessaloniki was on its way behind a Baldwin-built 2–10–0. The engine, delivered after the end of World War II, was now ending its days on the Greek main line. Swirling oil smoke up into the clear atmosphere, the Baldwin Decapod rolled by with only a handful of waggons: the scene is shown on plate 91. With no more steam scheduled for several hours, I wandered casually back towards the road, passing on the way a small overgrown coppice. Walking by, I inadvertently glanced into the trees and, to my astonishment, a rusted face of alarming familiarity silently gazed outwards. My initial reaction was one of abject dis-belief and I assumed the mirage to be a trick of branches, sunlight and leaves. I stopped and looked again. There was no doubt about it for, partly obscured by branches, I could discern the smokebox and chimney of a large Golsdorf engine from the old Austrian Empire.

After struggling through tangled scrub, I finally confronted the phantom, a 2–10–0 of the old Empire class 580. She had long since had an accident, her front being badly mutilated. Years ago, the wreck must have been pushed into this remote coppice and left to rust away. Remembering the Baldwin engine, I realized that within half an hour I had seen a couple of 2–10–0s very different in age and ancestry. But that rusted hulk, shown on plate 59, evoked within me poignant thoughts of the once vast Austrian Empire, which produced one of the world's most distinctive schools of steam locomotives.

The Austro-Hungarian Empire once covered a large area of Eastern Europe until the end of the First World War, when it was split up into several newly emergent countries. Much of its vast territory was both mountainous and sparsely populated, with the inevitable result that many lines abounded with steep gradients, sinuous curves and bridges of extremely light construction.

In 1891 locomotive history took a new course when Charles Golsdorf was appointed Chief Mechanical Engineer for the Austrian State Railway. His tenure of office lasted until 1916 and during this period he produced a prolific and imaginatively designed school of locomotives. It has been claimed that Golsdorf was responsible for some sixty different classes, though the figure is now generally regarded as being closer to fifty – a minor quibble when one con-siders that many of his designs were not only built in great numbers, but over many years as well, some as long as twenty-five years after his death. By the

Plate 58
Opposite: One of Europe's most exciting steam railways connects the Iron Mountain with Vordernberg in Austria. A standard gauge rack and adhesion Class 97 0–6–2T of late 19th-century design bursts dramatically out of a tunnel with a heavy iron ore train

82

early years of this century. Golsdorf had become well known to British locomotive students on account of his many papers published by the *Locomotive
Magazine*, his flair for publicity seemingly having led him to release some
account of every design he put into traffic. Though he had considerable influence
over the entire Empire, the Hungarian section was partly autonomous, as
witnessed by the use of main line Mallets from 1898 onwards and other solely
Hungarian engines.

The lightly built tracks and frail bridges forced Golsdorf to keep axle weights
under fourteen tons and this necessity, combined with a need for adequate
adhesion both for heavy hauling and surmounting gradients, led to his use of
many axles. The large boilers on his engines were necessary to provide adequate
power from the poor coal which predominated, and many of his creations
were truly enormous for their day. By 1897 he had produced his first 2–8–0
(170 class, plate 61), by 1900 an 0–10–0 (180 class, plate 60) and by 1911 had
even reached twelve-coupled. Obviously, such long wheelbases had to be
laterally flexible, and considerable side play was allowed. On his 180 class
0–10–0, drive was applied to the fourth axle in order to allow adequate side play
in the more strategically important third: extremely long piston rods were a
necessary feature of this arrangement. An even distribution of weight over the
axles was another requirement, so acute were the restrictions, and sometimes the

84

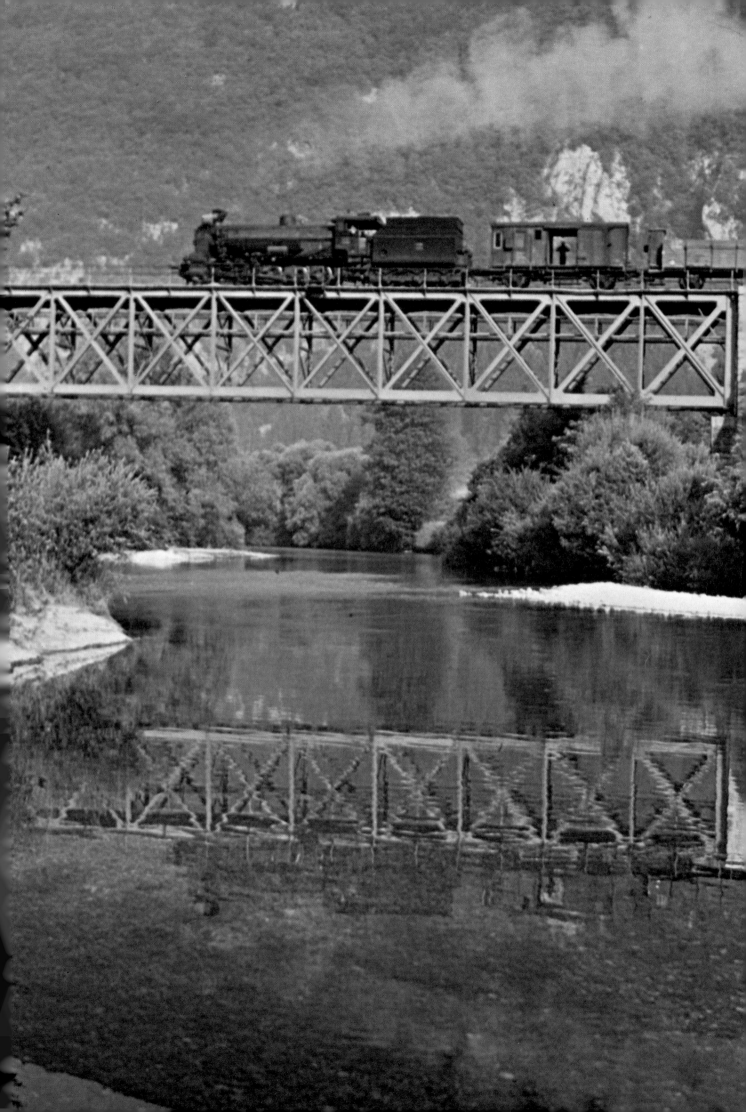

Plate 61
Equally famous as
Golsdorf's Class 180
0–10–0s are his Class 170
2–8–0s, first introduced in
1897. Originally 2-cylinder
compounds, these were also
built in simple form and
over the two variations
some 1,400 examples were
put into traffic. Both these
and the related 0–10–0s
have worked prolifically
throughout much of
Europe; here one crosses
the bridge at Maribor in
Yugoslavia

leading pony would be tucked back under the boiler, thus allowing a greater mass to be carried on the leading wheel. This gave some Austrian classes a characteristic appearance of jutting forward, as seen on plate 61.

Golsdorf was a great believer in compounding, and notable amongst his two-cylinder variants were classes 170/180, the latter having a low-pressure cylinder of no less than $33\frac{1}{2}''$ diameter. Many four-cylinder compounds were later built, though with the advent of super-heating a general transition was made towards simples, and the two classes mentioned above were also prolifically built in this form under the classifications 270 and 80.9 respectively.

Though of a functional nature, many excitingly ornate features distinguished Austrian engines. Especially notable were the double dome boilers with an external connecting pipe, a device intended to prevent the overflow of surging

86

water into the steam delivery pipe. Flared chimney backplates, stovepipe chimneys encircled with a lovely gold band, and smokebox doors constructed in two sections with a central opening were further typical hallmarks. In order to reduce weight, parts were pruned back, often accentuating an already stark appearance festooned with Westinghouse brake pumps and cylinders, spark-arresting meshes and water purifiers.

To many, the hideous and angular monsters thus created were both unique in appearance and totally alienated from any other line of development. In brief, they epitomized that foreign look so scathingly shunned by many English enthusiasts. Different though the Austrian engines were, they had tremendous personality, and a rare concession to traditional aesthetics was found in the gold band on their chimneys seen on plate 62. Here was a master touch, an understatement which signified more than a good deal of self-conscious designing.

When the great Empire was split up in 1919, the newly emergent countries of Czechoslovakia, Poland, Rumania and Yugoslavia all inherited Austrian design, and this formed, to a greater or lesser extent, an integral part of the steam motive power. In fact, building of Austrian classes continued in all four countries over some years, although Poland, Rumania and Yugoslavia eventually adopted a tendency towards German-orientated outlines and development, with Czechoslovakia and the much reduced Austria being the chief perpetuators of the old Golsdorf line. Czechoslovakia continued to build the 2–8–0/0–10–0 in simple form until the late 1920s, while the old 629 class, seen on plate 62, was continued right up until 1940. However, a very characteristic stud of Czech engines appeared over the inter-war period, culminating in the splendid 4–8–2s and 2–10–0 types built over the late 1940s/early 1950s, and in these the old Austrian strain is entirely absent.

The Second World War created another upheaval of motive power, and numerous ex-Austrian types were included in further territorial changes. Additionally, the influx to almost all East European countries of the ubiquitous German war engines, plates 75/80, along with the various American war types, such as the numerous S160s of plate 86, not only contrasted greatly with the Austrian types, but sometimes combined with them to form a partial backbone of heavy power until the end of steam. This was especially true in Austria itself and to a lesser extent Yugoslavia. Even today, the old distinctive Austrian strain is still evident and it comes as a breath of fresh air amid its German and American counterparts.

Austrian Empire Engines Today
Apart from their becoming increasingly thin on the ground, many Austrian survivors exist today in communist states, and are thus rather inaccessible. The most numerous examples come from the ranks of the simple classes 270 2–8–0, and 80.9 0–10–0. Sadly, some engines have lost their old smokebox doors, and run today with the conventional one casting type – an alteration which greatly nullifies the engines' delineation.

Golsdorf's products are most numerous today in the Slovenian region of Yugoslavia, where the two types just mentioned run as the Yugoslav State Railways (J.Z.) 25 and 28 class; also, some splendid two-cylinder 2–10–0s exist, classified '29' by the J.Z. Slovenia has been associated with these classes

as it was once part of Austria. Maribor is an important Slovenian steam centre, and from here the branch line to Bleiburg is operated by five 4–6–2Ts from the original Austrian Sudbahn Railway – class 629 introduced in 1913 (see plate 62). Now classified 18 by the J.Z., these engines came to Yugoslavia as part of a reparation package from Germany after the last war. The 25s can also be seen at Maribor – plate 61, and the two classes make an interesting comparison with the German-built 06 2–8–2s – plate 76, in representing two different streams of Yugoslavian steam traction. Plate 62 was made at Rûse, and here I discovered an old Golsdorf two-cylinder compound 2–6–0 tank engine busily shunting at a factory. She was typical of the many diminutive engines produced for remote branch line working throughout the old Empire territory: the engine still retained her double dome 'handle' boiler.

The Czechoslovakian State Railways (C.S.D.) cannot produce a greater variety of Austrian types, but they do have the 270 2–8–0s running under the guise of the C.S.D. 434.2 class; some retain their double dome 'handle' boilers. The 629s are also active under the classification 354.1; these, as with most 434.2s, are engines actually built in Czechoslovakia. Examples of the 270/80.9s might also be found in Hungary and Rumania.

In present-day Austria all Golsdorf products have long since disappeared on the Federal Railway – apart from the 629s which worked right up until 1974 as the 77 class, although their latter-day appearance was greatly altered by the acquisition of Giesl chimneys. However, the G.K.B. private coal railway at Graz still operates the 170 class compound 2–8–0.

One remarkable throw-back to the old Empire days survives in the Erzberg or Iron Mountain – see plate 58. This magnificent line carries iron ore from a 2,400 ft high mountain in the Austrian Alps down to Vordernberg; from here the ore continues along the electrified main line to the vast steel works at Donawitz. As it is adequately documented elsewhere, suffice it to say that the Iron Mountain line is one of the most exciting left in Europe today; especially in wintertime, when, after heavy snowfalls, the sight of two 97 class 0–6–2Ts contrasted against a white landscape and barking their way over a 1 in 14 rack section, pumping thick palls of rich brown exhaust up into the frosty air, offers a spectacle which few could fail to appreciate.

Over the years 1924–7 some 80.9 0–10–0s and 580 2–10–0s were supplied to Greece, where they became important freight engines. During my extensive visit to that country in 1973 none were found active, though derelict examples did exist in addition to the wreck shown on plate 59.

While writing this section on Austrian engines, I took a walk by the canal adjacent to the Midland main line in Leicestershire. Looking across two fields I could see the line emerging from a short cutting and standing out against a heavily cloud-flecked sky. I imagined the lean, elongated form of an Austrian 580 2–10–0 drawing a string of coal waggons with stately momentum through the soft countryside, her superb black livery glistening in the late afternoon sunlight. What a supreme apparition she made and how she enhanced the landscape; the curls of white, fluffy smoke from her gold banded chimney being snatched and twisted into oblivion by the breeze. There was a sight to rival the Midland 'Spinners' themselves! A blue bullet-nosed B.R. class 45 diesel swept southwards, shattering my illusion entirely, and leaving plate 59 as my sole memory of a beautiful breed of engines.

Plate 62
Opposite: An example of the stately Austrian Pacific tank engine – notice her smokebox doors and gold-banded chimney, both typically Austrian features. Initially produced for the Sudbahn Railway in 1913 as the 629 Class, these engines represent one of the most distinctive steam designs still extant in Europe today

BRITISH

Plate 63
Above: Wheels and
Walschaerts valve gear of a
British Railways Standard
Class 4MT 4–6–0

Plate 64
Right: The burnt up
remnants of Stanier
locomotive wheels on the
ground of the breakers'
yard

Plate 65
Opposite: The blackened
remains of man's industrial
past are still evident today
in the valleys of south
Wales. In the valley next to
the one depicted here the
world's first steam
locomotive was born in
1804. 175 years later the
end of the drama is enacted
as an Austerity 0–6–0ST
clanks its waggons back up
to the pit head at the
Hafod-rhy-nys colliery

Plate 66
Previous pages: Blaenavon
featured prominently in the
early days of the Industrial
Revolution, for this was a
time when its night skies
glowed red from countless
iron foundries fed by
attendant collieries and
limestone workings. All this
is now history and yet to
this day memories of those
halcyon days are still
evoked when Blaenavon's
two surviving Andrew
Barclay 0–4–0STs of
1919/20 vintage spin their
wheels with heavy trains
and momentarily light up
those night skies once more

Echoes of a Revolution

One of Britain's greatest technological gifts to mankind was the steam engine.
Its inception opened up trade and communication in a way that the canals
could never have done: it hastened the Industrial Revolution on its way and
changed the face of the entire globe. Iron and coal were the revolution's life-
blood and steam its heartbeat. Britain led the world.

The beginnings of that revolution came over two centuries ago when iron
foundries, fed by a multitude of collieries, ironstone mines and limestone workings, lit the night skies of South Wales. One great iron town was Blaenavon,
where the bitter-sweet intrigue of industrial clamour, albeit by exploited labour,
generated an energy and output which would be envied today. The minerals
were worked in their abundance: iron, coal, and later, steel becoming the life-
blood of those valleys and a pulse to Britain's prosperity. Steam engines pro-
liferated everywhere, flitting their way through colliery, ironstone mine, and
foundry: the one cohesive factor of the whole.

Today, the revolution's accent has changed, as has Britain's status, but the
memories linger on in those now blackened valleys of South Wales. One-and-
three-quarter centuries after its inception, Britain's steam age is played out –
a tantalizing three miles from where it was born in the next valley, as a Hunslet
Austerity clanks its waggons back up to the Hafod-rhy-nys colliery in the mid-
1970s: the grime-stained valleys, the derelict structures of man's industrial
past and the perennially grey skies adding an evocative foil to our memories of
those halcyon days – plate 65.

Now, Blaenavon is little better than a ghost town; obsessed by spirits of its past, it can bear little countenance to the present. Gone are the scores of labourers' hovels from the hillsides, gone are the taskmasters' mansions, and gone is the industry: the half-buried and weed-choked remains of the old foundries only serving to enslave Blaenavon to a world of historical introspection. Yet, in 1975, a harrowing testimonial to those epic days survives in one remote windswept colliery, jutting like a craving hand out of the emptiness of the blackened hillsides above the town. Here, survive two sixty-year-old Andrew Barclay saddle tanks which, diminutive as they are, have little difficulty in trundling the pitifully small output of present-day Blaenavon coal. Pictured on plate 66 are *Nora* on the left, and *Toto*, built in 1920 and 1919 respectively. *Nora* was daughter to a manager of the old Blaenavon Company and *Toto*, it is claimed, was the family's dog.

As the wind moans through Blaenavon's main street, one listens pensively for the sounds of an industrial cacophony centuries past; the ghosts are never far away. But better than Blaenavon's multitude of phantoms are the two Barclays going about their chores. When slipping with loaded trains they momentarily colour the night sky with fire and, transitory though this is, memories are conjured up of the fiery brilliance which once illumined the valley. But the town, inextricably bound with its past, feigns not to look.

British Engines for Home and Export

Over much of its history the British steam engine remained rather insular, both mechanically and aesthetically. For sheer symmetry of form and neat appearance, there has been little to compete with British engines, especially towards the end of the nineteenth century, when certain designers would permit no

Plate 67
Left: The last express passenger duties by steam in the land of its birth were enacted by the Bulleid Pacifics. Here one hastens a Waterloo – Bournemouth express through the Micheldever Tunnels in Hampshire

Plate 68
Bottom: In extreme contrast and yet no less British, is this Sharp Stewart 2–4–0 of 1880 design heading antiquated 4-wheeled coaching stock through the tropical landscape of Java. The 2–4–0 and Pacific types were at opposite ends of the evolutionary scale in express passenger engine design

mechanical improvements to their engines in case the appearance was impaired. It is little wonder that the British delineation has been widely envied and copied. Many attempts have been made to define why British engines should be so styled, some of the more blatant theories claiming it to be a result of high culture – the genteel mind being devoid of harsh and ill-mannered excursions – as against the rough-and-tumble of the world at large! Such a theory seems untenable when one considers the French and German schools, both born of long-developed cultures, yet their products were angular and unerringly mechanical, compared with Britain's. Obviously, such aspects as terrain, fuel availability and operational requirements all greatly influence the end product, and these aspects vary greatly throughout the world. Nevertheless, the unfailing ability of the British to produce handsome locomotives both for home and export is universally acknowledged: such a virtue cannot be totally divorced from the British temperament. Doubtless much inherent beauty in the designs can be attributed to the chief mechanical engineers of the old private companies, who traditionally have had supreme, if not actually autocratic, command over locomotive development, whereas in many countries design was done by a consortium of operating men and outside builders.

Britain's lead in railways, combined with her vast Empire, ensured boundless opportunities for the country's engineers and products. It was probably inevitable that her private locomotive builders should emulate, whether deliberately or psychologically, standards set by the main line companies, and thus the engines of pre-grouping Britain were destined to go to all corners of the world – not just within the Empire, vast as it was, but also to many lands devoid of political affinity with the great Britannic rule.

After the grouping of Britain's main line railways into four large companies, a less insular application to locomotive design prevailed, and developments from other parts of the world began to infiltrate with an unprecedented momentum. Ever-increasing competition from abroad meant that it was no longer good enough for locomotive exporters to mould heavily on the indigenous forms: engines had to be tailored to the specific conditions applicable in the recipient country. Foreign builders were offering some imaginative things and it became evident that what, for better or for worse, had been good for Britain, was not necessarily good for the rest of the world – although it often was. Thus those handsome and elegant machines with saturated steam boilers, slide valves, short travel valves, inside cylinders, low footplating decked with curved splashers, narrow between the frame fireboxes and often unsuitable bearing surfaces, which in their tens of thousands had rolled on to boats from British shores, began to change mechanically and thus in appearance too.

Throughout the last few decades, an interchange of world design ideas produced engines which were much more universal in their concept; many of the mixed traffic engines previously mentioned were in this category. Some obvious characteristics absorbed from world trends were: superheating, long travel piston valves, outside cylinders with better designed steam passages and wide fireboxes to help consume the indifferent coal found in so many countries. Towards the end, the traditional American concept of two outside-cylinder simples was almost exclusively adopted, along with such latter-day American refinements as self-cleaning fireboxes, rocking grates, hopper ash pans, steel fireboxes and cast steel beds, while roller bearings on all axles or manganese

Plate 70
Deep in the Tanzanian interior splendid British-looking engines work on to this day. The engine here is an East African Railways Class 25 2–8–2 built in 1926 by the Vulcan Foundry, Newton le Willows, Lancs. She was originally built for the British territory of Tanganyika as that country's MK Class

steel on both plain bearings and rubbing surfaces were a further refinement.

Clearly it is impossible to give exhaustive treatment to so wide a subject, but possibly a superficial pattern might be seen by following the various accompanying plates. Plate 68 shows a typical mid-nineteenth-century export which would have looked equally at home in England. Moving to a slightly later period, plate 72 shows an engine highly representative of early twentieth-century design – look at her tall, slender chimney, curved steam pipes, rounded cylinders, low footplating and incredibly neat delineation. More modern yet, though still unmistakably moulded, are the engines on plates 69 and 71: a 4–8–2 and a Pacific respectively, both from the design period immediately prior to 1920. However, plate 31 shows a confrontation between the South African classes 15AR and 15CA – both 4–8–2s; the former of 1919 was among the last of the pure British engines to be supplied to the Republic, while the latter is the first representative of the big American engines, introduced in 1925. These were to have a considerable effect on future steam building in South Africa.

If the subject of plate 70 had been seen at Derby in the mid-1920s she would have evoked little comment; neither would the engine on plate 73 have done a few years later. A different contrast comes on plate 74, which shows a 1952-built 2–8–2 from Bagnalls of Stafford adorned with Giesl chimney, high footplating and feed water heater, while in the North British-built Condenser on plate 35 the metamorphosis is complete, wherein a fusion of design practices has produced a thoroughly modern engine bearing little evidence of its parentage.

Finally, a few pictures depicting the last of British main line power at home: plate 67 shows an express on Britain's last all steam main line from London

(Waterloo) to Bournemouth. This route remained steam operated until 1967. Meanwhile, further north, a line of Stanier 8F 2–8–0s lies derelict inside their closed depot at Patricroft in Manchester, the broken slats of the old shed roof allowing the sunshine to filter in and transform the entire scene into mottled effect – plate 30. Plate 64 is a companion to this scene, showing the wheels of these very engines as they were three months later, having been burned up at Cohen's scrapyard in Kettering: the very end of main line steam in the land of its origin.

British Engines Today

The ex-L.M.S. Stanier 8F 2–8–0s, introduced in 1934, were once important main line freight engines in Britain. Additionally built for the last war effort, many graduated abroad and although most returned home, some were retained by Turkey where they still survive on light duties. These are the most notable 'British School' engines left in Europe today and are shown by plate 30. Others, also left behind by the war, still soldier on in Iraq.

Some early twentieth-century British 'classics' remain in India. Over many years India's various railway companies all had their own designs and liveries and were so utterly British in application, that by the turn of the century India's railways were little different from those of England. So diverse had the designs become, that the British Engineering Standards Association in London designed a range of standard types for the sub-continent. Moulded in the best of home traditions, they were comprised of 0–6–0s, 4–4–0s, 2–6–4Ts, 4–4–2s, 4–6–0s and 2–8–0s. Although more recent years have seen India's railways follow a different course, some precious remnants of the imperial reign still operate in ever dwindling numbers and 4–4–0, 4–6–0, 4–6–2 passenger engines, along with 0–6–0 and 2–8–0 goods engines remain active.

The 4–4–0s are inside-cylinder thoroughbreds with big splashers; their 'H' class 4–6–0 'Mail Engines' would look much more at home between Marylebone and Manchester, while some of the 0–6–0s look unnervingly like the old Great Central J11 'Pom Poms'. A report on Indian steam in the *Continental Railway Journal* said: 'Mechanical maintenance is in the good old English tradition and engines tend to wheeze, clank and bang their way along in a manner sufficient to make an ex L.M.S. or L.N.E.R. man green with homesickness. Whistles are of the squeaky L.N.E.R. type, though some engines have a mellow Great Western or Great Central tone.' Little more need be added – except where to obtain the cheapest air fares! Neighbouring Pakistan is also blessed with inside-cylinder 4–4–0s and 0–6–0 goods engines – both of applaudable vintage.

Possibly the only country to compare with India for the 'classic' engines is Argentina. Though not part of the Empire, Argentina's railways were constructed by British engineers and largely controlled by British capital – a situation which existed until as recently as 1943, when the Argentine Government bought out all remaining British interests and took their railways under state control. Apart from the usual range of lovely designs, Argentina's engines once burned imported Welsh coal in their narrow fireboxes! The usual breeds can still be found: shapely Pacifics without so much as a nut out of place, 4–6–0s with a tantalizing hint of the Great Central 'Fish Engines' about them, and Scottish built 0–6–0 freight engines of the 1880s – to name but three mouth-watering examples.

After the tragic demise of steam in Australia, it is the African continent which now provides the greatest evidence of its colonial past. Even technologically developed South Africa still has some old class 6 4–6–0s introduced in 1892, which bear more than a superficial resemblance to the old Great North of Scotland D40 4–4–0s of the same period. Look at plate 71. The engine possesses more than a hint of Great Western – albeit it quite accidentally – while the overall scene could have been anywhere in England between 1910 and 1960! Other lovely manifestations are seen in the 12A type on plate 69 and the tank engines on plates 45 and 73: all could have come from nowhere but Britain, and I regard them as some of the best-looking industrial engines left in the world.

Among the other old British colonies of Africa; Kenya, Tanzania and Uganda are of special interest. The engines in these countries are painted in delightful shades of maroon – which vary considerably from engine to engine, as frequent shortages of paint necessitate the admixture of other colours. They also carry names and are among the very last remnants of this simple, but profoundly satisfying, practice: if Britain had led in nothing else, she would certainly have taken the laurels for named engines. Over one-and-a-half centuries, British locomotive names were both imaginative and tastefully chosen and, apart from heightening an engine's individuality, they added tremendous interest, particularly when the same name was passed down through various generations of different locomotives, as on the old L.N.W.R. and into the subsequent L.M.S. Decorated in their maroon liveries (which range from bright scarlet to burnt umber), East Africa's named engines offer a splendid relief against all other steam fleets. Certainly the practice of naming main line engines spread in various degrees throughout the British Empire, as of course did colourful

Plate 72
A priceless late 19th-century British export with more than a hint of Victoriana is this ex-South African Railways Class 7A 4–8–0 currently enjoying a new lease of life over the colliery metals of the Transvaal. She came from Sharp Stewart of Glasgow in 1896

103

liveries, but little evidence of either exists today, East Africa being the great exception. The 'Mountain' class Garratts seen on plate 22 have little extension plates added to their names indicating the height in feet of the particular mountain commemorated: this is similar, in principle, to the badges and various other emblems traditionally affixed to the names of many British engines.

The class 25 2–8–2 shown by plate 70 is done out in maroon and she looks ingeniously L.M.S., but the related 26 class on plate 74 has her thoroughbred lineaments almost totally disguised, and looks both British and German. It needs few embellishments indeed to transform radically the atmosphere of any engine.

The almost complete metamorphosis undergone by the latter-day British steam exports is personified by the mixed traffic 'Tribal' engines, currently responsible for most of the steam traffic in East Africa today. The three very similar designs total 103 engines which were built by North British of Glasgow and the Vulcan Foundry in Lancashire. Named after the East African tribes, these engines possess many modern sophistications, and the old British look has been sublimated to produce a highly efficient design prepared in the un-biased light of world experience.

British Nostalgia

Recent British governments have perpetually warned the nation about making an insufficient effort and living beyond its means. Soaring inflation, low exports and a poor balance of payments all weigh heavily against the nation's future

Plate 73
A North British 4–8–2T trundles a 900-ton coal train through the golden grasses of the Transvaal winter as part of her duties for the Greenside Colliery, Witbank. The figure sitting on the engine's buffer beam is the 'sandboy', an African labourer whose job it is to throw sand on to the rails ahead of the engine to reduce slipping when heavy trains are hauled over steep gradients

prosperity and well-being. These ills, we are told, are portents for the blackest time in Britain since the last war, and yet the nation seems ill-equipped either to understand the situation or to provide a remedy; the backbone of British morale is being severely weakened. That magical spirit which is said to emerge in times of heavy stress seems lost in all the confusion. In fact, this period seems more characterized by low morale than any economic hardship, with the nation seeking solace from its insecurity in higher monetary rewards. Many, though desperately wishing Britain to retain her sovereignty, miserably look towards Europe and the Common Market as the one saving grace, clutching our neighbour's hand lest we fall.

During the winter of 1974–5, I visited societies of varying kinds in the old locomotive building towns of Manchester and Leeds. Here, as in so many industrial places, that old-time pride and enterprise is yearned for with a poignancy which would have been unimaginable a mere ten years ago. Obviously the steam locomotive is but one symbol of our national pride, but in those towns especially, they still talk of the machines which in their thousands rolled from the great workshops bound for all areas of the world. When it is revealed that many shipped out as long ago as Victoria's reign are still active, then the nostalgia, especially among the older generations, can be almost embarrassing.

'Where has our enterprise, workmanship and spirit gone?' they ask. 'We have since become the poor nation of Europe.' The steam engine is now seen by many Britons as being one of the pillars of better times gone by – both industrially and socially. If only for this reason, it has a living place in our society today.

Plate 74
This wonderful British thoroughbred in 'improvised' East African Railways 'maroon' livery is greatly disguised by the Giesl chimney, Feed water heater and Westinghouse brake pumps and cylinders. She is the last survivor of the old Tanganyika Railways ML Class and retained for use today along the desolate 130-mile long Mpanda branch set amid wild scrub country – on account of her 9¾-ton axle loading. She was built by Bagnalls of Stafford in 1952

German Engines

Over the last decade, German engines have assumed tremendous popularity; their powerful, rugged and masculine appearance appealing to the modeller and lineside enthusiast alike. With steam still active in both political sectors, many people visit Germany each year to see the engines in action. German design influences constitute the predominant strain in European steam today.

After the First World War Germany's railways were unified in 1920 to form the Reichsbahn, and this absorbed such historic state railways as the Bavarian, Baden, Saxony, Wurttemberg and Prussian. From these and other systems, the Reichsbahn inherited almost three hundred different classes and this precipitated one of the most monumental and effective standardization schemes in locomotive history. Based on the fine locomotive tradition of Prussia, the Reichsbahn produced twenty-nine standard types, some with the Prussian outline faithfully perpetuated. As might be imagined, the Prussians had already produced some epic-making standards of their own, with a high interchangeability of boilers and many other parts. A notable example was the old P8 4–6–0 built between 1906 and 1924. These engines had superheaters and piston valves and were the antecedents of the mixed traffic 4–6–0 later adopted by many countries: over three thousand were built. The G8 0–8–0 and G10 0–10–0 – which had an interchangeable boiler with the P8, were also built in their thousands, while the three-cylinder G12 2–10–0 of 1915 was the direct fore-runner of many later Reichsbahn 2–10–0s.

The Reichsbahn standards form the basis of German steam today. Predominant were the Pacific, 2–8–2 and 2–10–0, which although accounting for only

Plate 75
Today the most prolific and widely distributed European steam engines are the German 2–10–0 *Kriegslok* – war engines of World War II which are still operating in considerable numbers throughout many countries. Here are two of them in the gloom of Vienna (Nord) depot

Plate 76
A typical German export
during the inter-war period
were the Class 06 2–8–2s
for Yugoslavia – one of
that country's few truly
indigenous designs. Here
one is seen leaving Maribor
in Slovenia with a
passenger train for Cacovec

eleven designs, constituted the greatest majority in individual units – 13,000 engines against some 1,500 of other types. Simple traction was adopted in accordance with latter-day Prussian practice, and other traits were long-travel valves, bar frames, high boilers with very short chimneys, low mountings and large sideplate windshields. Such distinctive outlines were further emphasized by two domes, boiler mounted sandboxes and an angular cab design to provide a roomy interior.

The engines were orderly in overall appearance, but their pattern of iron-work, although controlled and ruggedly satisfying, never approached the symmetry of the British. There is, however, a 'racial' similarity between the two schools, particularly in twentieth-century design, and one might almost say that the German aspect represented the masculine gender and the British its feminine counterpart.

The Reichsbahn locomotives greatly influenced other countries, especially Poland, Bulgaria, Turkey and Yugoslavia, where the modern designs were either copies with slight embellishments, or new designs prepared within the basic format. A policy of extensive standardization of parts, especially boilers, was adopted by Bulgaria and Yugoslavia. However, the spread of German types was not restricted solely to design influences and exports. Two great wars ensured a tremendous distribution of German locomotives: Prussian engines during the First World War and Prussian/Reichsbahn types during the Second. Military occupation by Germany of vast areas of Europe's land mass guaranteed this prolific distribution for her engines, while her subsequent defeat

in both wars meant not only a loss of territory, but the demand of large reparation packages to the aggrieved countries. Thus, by fair means or foul, the great German classes colonized Europe and in many cases proved superior to the indigenous engines of the recipient countries. Even in Russia, where the rail gauge is 5′ 0″, many captured German 2–10–0s saw service in modified form; and years after the war some were 'de-modified' – and sold for further service in Russian satellite states. So the general excellence of German design spread, and its acceptance enabled neighbouring countries to Teutonize, so providing Germany with a semi-captive market for peacetime exports.

Plate 77
Opposite: Against a brilliant sunrise a German Federal Railway Class 44 3-cylinder 2–10–0 labours its way through the Mosel Valley at the head of a heavy freight train

Now, thirty years after the war, the most common European steam engine is the classic *Kriegslok* (war engine) 2–10–0 – plates 75, 81. These were simply an austerity version of the standard Reichsbahn 50 class 2–10–0 of 1938. Over six thousand of these shorn-down and barren variants were built to follow Hitler's armies, and they have since become a standard European mixed traffic class. Another great Reichsbahn standard which possesses considerable distinction is the 44 class three-cylinder 2–10–0 of 1926 – plate 77: descended from the Prussian G12, almost 1,800 had been put into traffic by 1944.

After World War II the 'standards' were split between a divided Germany and little was added in the way of new design before the decision came to abandon steam altogether. One notable exception was the 23 class 2–6–2s – a design intended to replace the multitude of ageing Prussian P8s: both East and West sectors built over a hundred engines to a similar basic design.

German Engines Today

West Germany's line between Rheine and Emden has become a world-famous Mecca for enthusiasts. Pacifics, and big freight engines roar past in epic fashion on a route which retains all the atmosphere of a busy steam-worked main line. The old Prussian engines are now gone from West Germany, though a few still soldier on in the Eastern sector. At the beginning of 1975 West Germany had over 600 steam engines of five basic classes: Reichsbahn Pacifics, 2–8–2s and 2–10–0s formed the great majority, but a few post-war 2–6–2s were running as well. By comparison, East Germany had about 2,000 active engines covering fifteen different classes.

Poland, much of whose territory was once in Prussia, took many of the old designs after World War I and continued building them. The P8, G8 and G10 became national standards, and all three remain in evidence today alongside Poland's more modern Germanic power. Yugoslavia's only indigenous steam engines are the 110 Pacifics, 2–8–2s and 2–10–0s built by Schwartzkopffs in 1930; all have identical boilers – plate 76.

Turkey retains a fascinating array of German steam. She has both Prussian and Reichsbahn types in addition to some export designs specially built for the country's difficult terrain, including 4–8–0s and 2–10–2s. Turkey is also re-nowned for still operating German, British and American war engines. Even in Austria, the ubiquitous *Kriegslok* now forms a large part of the country's ever diminishing steam fleet – witness plates 75 and 80, and not one Golsdorf engine survives on the main lines. Prussian G8s can be found in the Lebanon, while Syria can boast both G8s and G10s – the latter still working the Taurus Express. Who would ever believe that a 1910-built Prussian 0–10–0 freight engine would be hauling an international express train in 1975!

109

The prolific export trade conducted by such renowned German builders as Henschel, Orenstein & Koppel, Hartmann, Krupp, Maffei, Esslingen, Borsig, Hanomag and Schwartzkopffs gave their country's engines, whether main line or industrial, a world-wide distribution: some were built to the national pattern and some to the specified design of the purchaser. They remain widespread today particularly in Asia, China, South America, South Africa and Portugal. Plate 76 depicts a surviving glimpse of nineteenth-century German steam.

Plate 79
Opposite: Prussian locomotive traditions formed the major constituent of Germany's railways when they were unified in 1920. Here another Prussian remnant in the shape of a Class T18 4–6–4T runs today in the Black Forest area as the German Federal Railways 078 Class. Over 500 of these engines were built from 1912 onwards

Plate 80
Left: Another German war engine, this one partially disguised by a Giesl chimney, charges out of a loop on the Linz–Summerau line in north-western Austria. These 2–10–0s are classified 52 by the Austrian Federal Railways

Plate 81
Following pages: Hitler war engine ambles through the quiet Slovenian landscape under the guise of one of the Yugoslav State Railway's Class 33 2–10–0s

American Traditions

The lack of cultural history in the new world was greatly offset by the romance of its pioneering days. The great railroads cut across the new continent with locomotives of such boldness and flamboyance in design as had never been seen before, and the classic American 4-4-0 of those early days was to become as legendary as Huckleberry Finn. Today, well over a century later, their glorious shape and sonorously moaning chime whistles can still be found amid the closing ranks of world steam.

Directors of 'Wild Western' films seldom miss an opportunity to present vivid scenes of some racy-looking, balloon-stacked 4-4-0, belching black smoke and vigorously clanging a golden bell on its boiler top. Walt Disney's epic film *The Great Locomotive Chase*, apart from being one of the finest triumphs in capturing steam's appeal ever made in any of the art forms, clearly reveals the passion with which he regarded these machines. In fact, the endearing shape of early American engines, with their magnetism and continuity of style, captured the world's imagination. Those locomotives epitomized freedom: not just because they rolled over the wide, open spaces of a new country, but in the generous permissiveness and unashamed flamboyance of their design as well. The superlatives used elsewhere in this volume to describe other schools of design must not detract from the one presently under discussion; the American school was as exciting as it was unique.

Its hallmarks are legion: ornate headlamps backed by enormous spark-arresting smokestacks, cow-catchers, warning bells, rounded cylinder casings topped by a rectangular valve casing, two domes – the leading one for sand, the second for steam collection – and, right from the early embryo forms, raised footplating. One of the loveliest features is the eight-wheel bogie tenders – traditionally lettered out as in plate 25. Wooden cabs were another early feature. Everywhere the stays are frequently commented upon, which connect the smokebox with the engine's front framing, as American engines have each outside cylinder cast with half the smokebox-saddle – the latter being bolted down the centre – and the familiar stays are necessary to strengthen the forward frame extension. This method of construction provides an excellent anchorage against piston thrusts.

On a continent so large and underdeveloped, loading gauge restrictions were, from the outset, less severe than in Europe, thus generously proportioned engines were possible and quickly found to be essential. Furthermore, long runs over new and sparsely populated territory inescapably meant an indifferent track quality. The 4-4-0 possessed greater stability than the early Singles and 2-4-0s and was widely adopted, first appearing in the 1830s. This was some fifty years before the type became widespread elsewhere, and it is still referred to as the 'American'.

It is interesting to see how far ahead America was in the evolutionary development of motive power: a 4-6-0 had appeared by the mid-nineteenth century and a 2-8-0 by 1866, while by 1890 both Pacifics and Atlantics were in existence. All of these were years ahead of their application in other areas of the world. Another American type, the 2-6-0 'Mogul', has been adopted for innumerable designs, both home and export.

With few notable exceptions, the basis of American design was for simple, robust construction, and two outside-cylinder 'simple' engines have been a principal form throughout. In looking back over the gamut of world locomotive history, one thinks of the innumerable incursions into various cylinder combinations and positions, with both simple and compound expansion, and all the attendant theories thereupon, yet in the end, the world's last steam engines were to reconcile themselves to the traditional American concept.

Insular though American practice was, its traditional characteristics were to spread far and wide, either by design influences or physically exported engines, especially over the later years of steam. The use of bar frames became widespread, along with its subsequent derivative, the integral cast steel bed, which avoided the traditional bolting up of bar frames by incorporating in one complete casting the cylinders, smokebox saddle, side frames and cross members. Wide fireboxes which fanned outwards over the trailing axles provided tremendous heat and good consumption of poorer fuels, and were often well received in export packages as against the narrow British box placed between the frames. The use of steel fireboxes eventually found a widespread application, as did mechanical stokers. High pitched boilers and maximum accessibility to working parts have always been characteristic of American design, both practically and visually.

With so much energy expended upon high utilization and trouble-free running over great distances, it is understandable that many improvements of detail should derive from America, and such aspects as self-cleaning smokeboxes, hopper ash pans, rocking grates and roller bearings all made their mark over the closing years of world steam construction.

Possibly the two most influential European aspects to pervade the fortress of American steam both occurred in the early years of this century; one was superheating, the other, the Mallet articulated. Within thirty-five years the Mallet had flowered into the biggest steam engines the world will ever know – the 131 ft long, 520 ton Union Pacific 4–8–8–4 'Big Boys'. These, and others like them, roared their way across the great plains like their dinosaur counterparts millions of years previously and took their place alongside the more general latter-day 4–6–4/4–8–4 giants, in what was the world's greatest steam spectacle until it was swept away from its homeland by a tornado of dieselization.

Though retaining a separate identity throughout, American steam, both for home and export, achieved, in very general terms, two extremes of design. The old nineteenth-century style, though giving way on the one hand into a succession of bigger and more elongated engines leading to the final flowering just mentioned, did retain much of its overall delineation in the smaller designs and was preserved, albeit in a refined form, right up until the end: evidence of both extremes still exists and is discussed in the following pages.

Built as recently as fifty years ago, the subject of plates 82 and 89 retains much of the older strain and atmosphere. An intermediate period design is seen on plate 92, while the modern look, although in scaled down form, is shown by plates 86 and 90. Only in Java can big main line Mallets be found and one is seen on plate 84: she is a Swiss built descendant of an original Alco four-cylinder compound design of 1916.

It is perhaps surprising that the family likeness of American engines has survived so well in a country where private railway companies were numerous and a diverse export market industriously pursued. But the bulk of American engines, especially during the present century, came from the 'big three' builders:

Baldwins, Alco and Lima. Accordingly, over the years, progressive designs were improvised and embellished to suit the needs of recipient concerns, whether in America or overseas, and thus something of a conveyor belt system operated, enabling engines to be built quickly and cheaply. The rugged and chunky exteriors inevitably produced did little to deter purchasers from engines whose prime virtues lay in an unfailing ability to steam and a reliable utilization in service. Hardly any of America's railroads built their own engines, and a particular design or class would be born from consultations between the company and the big builders.

The opposite applied in England. Apart from every major railway company designing and building most of its own engines, the design dictates lay almost exclusively in the chief mechanical engineer's hands, and successive C.M.E.s often completely reversed the policy of their predecessors! Certainly the British school was discernible for all its inherent variety, but it inevitably lacked the remarkably consolidated unity of American traditions in general.

Today locomotive students seem to love the two great extremes of American engines equally well. The old school was vividly recalled for me one foggy evening in Smethwick during the winter of 1975 when, in an antique shop, I discovered a coloured revolving table lamp depicting a mid nineteenth-century 4–4–0. I am not normally one for bric-à-brac, but this was irresistible and I secured it for a very modest sum. Now in pride of place, the 4–4–0 blazons her way across the dining room. She is seen highballing across the prairie against a mountain backdrop at around 50 m.p.h. (nearest possible guess), colouring a bluish-grey evening sky with flame and black smoke – the glow from her

Plate 83: A locomotive's chimney is an important focal point: Dragon 7's appearance is totally altered by the slender 'stovepipe' chimney, for she is otherwise identical with Dragon 4 opposite. The ornate spark-arresting chimneys are fitted to the bagasse burners, but early in the milling season some engines burn oil until sufficient new season's bagasse has been dried and baled up. Dragon 7 came from Baldwin in 1928

decorative headlamp eating into the gloom ahead. Though hybridized in design, her parentage is faithful down to the very last details.

American Engines Today

If big American steam were active today, it would provide the greatest attraction of all for the ever increasing number of globe-trotting enthusiasts anxious to witness the last embers of the steam age. Even Mexico, which perpetuated the tradition some years after its disappearance from the homeland, has now completely succumbed to the diesel.

Where does the keen rail fan go in order to experience something of American traditions? Earlier it was said that the tradition had two distinct extremes: the flamboyant machines of the halcyon days and the ultimate giants of the mid twentieth century. Irretrievably the true giants are gone: no big Mallets and no mighty 4–6–4/4–8–4s, for such types handled the very top duties and today, almost all over the world, such work is now bestowed upon modern motive power. Paradoxically, it is at the other end of the scale that the most exciting evidence exists, and exist it certainly does, for in Peru an 1872-built 'Rogers' 4–4–0 still performs on a sugar wharf. Over a century old, she is the classic locomotive of the early school. Complete with wooden cab, this veteran, along with a similar Baldwin-built sister of 1910, works at Puerto Eten. Still discussing classic centenarians, the Campos Sugar Plantation of Brazil sports, among numerous other delights, an 1876-built Baldwin 4–4–0. For all the vintage and

117

Plate 84
Plate 84
Only on Java can the main-
line Mallet be found today,
and although this giant
Class CC50 2–6–6–0 was
built in Switzerland in 1927
she does represent the last
of the big Mallet tradition
which so dramatically
characterized much latter-
day steam development in
America

nostalgia conjured up by my revolving table lamp, these three antiquities look
even better!

Also in Brazil some splendid 2–6–6–2 Mallets are active on the Teresa Cristina
coal railway. Built by Baldwins in 1950, these metre-gauge engines provide
tender Mallets on big hauls and, although not matching the old main line performances, are aesthetic in being four-cylinder simples – as were the biggest
American Mallets: these Teresa Cristina engines are probably the last simple
Mallets in operation today. South America in general is now the best region for
the 'big three's' products: Brazil has a splendid range of American-built power;
Equador offers Baldwin 2–8–0s, and Baldwin Mikados perform in Chile – to
mention superficially but a few.

If a lack of vintage can be offset by decrepitude, then the Philippine Islands
harbour some of the finest pieces of Americana left today. As an example, look
at plate 89; the prolific imaginations of Disney and Emmett combined could
scarcely match this! Apart from oozing the old-time flavour, she contrasts her
rusted smokebox and balloon-stack against a silver boiler and green cab; if this
be insufficient, then her red and yellow tender almost completes the colour
spectrum. Notice the tender sides sheeted up with corrugated iron and held in
place with bits of string and wire; this is to keep the bagasse dry should typhoons
come. Her boiler top is decked with two enormous barrels, echoed by a third one
situated on the running plate: these contain sand to prevent the engine from
slipping on wet rails and, lest the sand should get wet, the barrels are also
protected by sheets of corrugated iron. A further adornment, albeit temporary,
is the large cube of bagasse inadvertently left on the running plate. One wonders

if the Wild West, for all its alleged wildness, ever produced such rustic delights as this. She works for the Ma-Ao Sugar Central on Negros – a highly efficient concern with a large annual milling tonnage. However, the Hawaiian-Philippine Co., some thirty miles away, gives a better impression with its immaculate Baldwin 'Dragons', and plates 56, 82 and 83 show these gorgeous engines freshly painted and overhauled for the milling season in October 1974.

In marked contrast, British and European enthusiasts have become familiar with the American war engines still active in Europe, these having been incorporated into the rosters of many countries after the last war. The best known are Major Marsh's S160 2–8–0s shown on plate 86. These were specially built to the British loading gauge and some actually worked in England in 1942 before going abroad to follow the Allied Armies. When the war ended, the S160s were avidly snatched up as cheap power, especially by the communist countries, and apart from being active in Greece, they still abound in Czechoslovakia, Poland, Hungary and Turkey. Nearly two thousand S160s were built by Baldwin, Alco and Lima, and they are a classic twentieth-century design. Widespread in distribution, S160s were the last steam locomotives to run in Alaska. Amongst other American war engines to survive in Europe are some ex-Army Transportation Corps 0–6–0 shunting tanks. Another type is the Baldwin 2–10–0, seen on plates 90 and 91: the engines clearly reveal their modern American delineation. Also, after the war, Poland took 100 twenty-ton axle-load 2–10–0s from Baldwin, Alco and Lima and she continued building similar engines up until 1957. Classified Ty246 and Ty51, these form an important part of Polish steam today.

Russia took large quantities of big American engines since physical conditions

Plate 85
The influence of big American engines came to South Africa in 1925 with the introduction of these large 4–8–2s from Baldwin and Alco. These Transatlantics were destined to have a considerable influence on future South African locomotive policy. Here one returns coal empties from Pretoria to the Witbank coalfield

in the two countries are similar. The presence of so many American engines in Russia had a considerable influence on later Russian design, since Soviet policy was one of incredible standardization, with successful designs being multiplied in their thousands. Both American and home-built engines exist side by side in Russia today. Neighbouring Finland, whose 5' 0" gauge conformed with that of Russia, produced a very neat family of engines noted for their simplicity. Many of the country's early designs were American built, and the modern 2–8–0 shown by plate 92 is descended from an earlier Baldwin class: she retains something of the original flavour.

Much American design influence can be dated from about 1920 onwards, when several important countries began to orientate towards transatlantic thinking, without necessarily taking many engines actually built in America. A dramatic example was Australia, where up to 1920 British engines predominated, but a general Americanizing of design led to a complete transformation, including the adoption of cast steel beds.

The old English atmosphere which continued on India's railways until well into the twentieth century is widely known, but in spite of England's semi-captive market, conditions on the sub-continent led to the adoption of some American influences. Although wide firebox engines were exported from Britain to India during the inter-war period, owing to the poor quality of Indian coal, certain dissatisfactions caused the now famous WP Pacific to be prepared by collaboration with Baldwins. This design incorporated many American features, including steel fireboxes and bar frames, as indeed did the related WG 2–8–2. These classes represented a considerable departure from tradition both mechanically and visually. Noted for both steaming capacity and reliability, the two types were destined to become India's most numerous steam classes.

South Africa's main lines have recently become world renowned for their epic steam performances, and the dramas currently enacted in lifting big tonnages over the undulating veld have frequently been likened to those of the great American days. In 1925 South Africa took some American 4–8–2s and Pacifics which were destined to have a revolutionary effect on future conventional power in the Republic. These were big American engines and one must not be misled by the Republic's 3' 6" gauge, since considerable tolerances are allowed both in height and width; of course the narrow wheelbase causes a high centre of gravity and thus speed is restricted to 50 m.p.h. The 4–8–2s are known as 'Big Bills' and, as plate 85 shows, they still pound tonnage across the Transvaal today. From these insurgents came the mechanically stoked 15F/23 class 4–8–2s, engines which now work the trunk routes across the Orange Free State – plate 33. When double-headed, these engines operate 2,000-ton trains at a steady 35–40 m.p.h. notwithstanding adverse gradients of 1 in 100; their bellowing exhausts and shrill chime whistles constitute big railroading at its best.

The foregoing are but a few examples of America's widespread influence both in physical exports and design practice and I hope some basis for further study has been provided. Once again, China is the one notable omission, but she is known to have taken large quantities of American locomotives and doubtless many are still in action: perhaps in more enlightened times it might be possible for us to see them.

Plate 86
Opposite: A classic American War Austerity: one of Major Marsh's celebrated S160 2–8–0s built from 1942 onwards for wartime service in Britain and Europe. Despite their cheap construction many survive today, notably in Greece where they are now classified Theta-Gamma. Here one lifts a heavy goods bound for Athens down the main line from Levadia. The sheet of flame erupting from the firebox is the result of rather over-ambitious oil-firing

The Insular Lumber Company

I have left until last an account of an almost unbelievable railway whose engines are, in my opinion, the most astonishing left in the world today. I suppose all of us foster dreams of idyllic situations which, if fulfilled, greatly enrich us in spirit, but seldom have I known such perfection as my experiences at the Insular Lumber Co, Fabrica on Negros Island in the Philippines.

About four years ago, research threw up the existence of this line. The story went that a decrepit and ancient Baldwin Compound Mallet and a bunch of even older two- and three-truck Shays, were operating a remote logging line on a small Philippine island. The engines, I was told, looked like nothing on earth, and apparently the Mallet worked a rickety, disjointed 3′ 6″ gauge line which carried teak from mountain forests down to the sawmills at sea level, the Shays working at the mountain loading area and also around the sawmills. My imagination ran riot, but when I actually saw pictures of the engines I just could not believe my eyes. So began the dream and my avowed intention to reach Negros before they all disappeared. Some months later I heard that the Mallet was virtually unworkable and that the company was scheduled to close the operation down as the mountain forests were almost cut out.

Having eventually formulated a date for my visit, I intrepidly sent a letter to the company, since they had no telephone, asking them to confirm that they would still be operating. The weeks drew by and no response came: were they just not going to bother sending a reply or had they already closed down? My last information had been two years previously and since then the whole set-up could have disappeared and the engines been burned up by a local scrap merchant. I began to convince myself that my dream was too good to be true; but then a letter arrived. It looked thin and without much content and, as I

Plate 87
Further American influence in South Africa is found in these lightly axle-loaded Class 24 2–8–4s built by the North British in 1949/50 and incorporating such transatlantic features as high-footplating, American-built frames and cylinders in one steel casting and Vanderbilt tenders for long-range operation in dry areas. This 24 has the Indian Ocean for a backdrop as she works between George and Knysna in the Cape Province

tore it open I mentally read the words 'We regret to inform you that logging operations ceased eighteen months ago'. But this was the fulfilment of a dream, for there, in one line, ran the words I will never forget, 'Our Mallet and Shay locomotives will be operating at the time of your proposed visit and we shall be pleased to welcome you to Fabrica.'

Ten days later I was there, among the world's most incredible steam survivors and on an island in which the charm and hospitality of the inhabitants is devastating. Philippine hospitality, they told me, is the best on earth – and they were right. The quiet, unhurried pace of living and simple but ingenious life, combined with a beautiful landscape, coastline and climate, utterly captivated me. There are countless thousands of people throughout the world who love both steam engines and good company: to them I can only say, you should have been born on Negros.

But the engines – how can I describe them? Pride of the line was certainly the Mallet; I had let my imagination run wild about her, but when she first confronted me I was literally spellbound. Apart from being almost twice as large as I had imagined, she had a marvellously faded green livery lined in red, while her eight-wheeled bogie tender was black with white lettering which announced: INSULAR LUMBER CO; lesser scripts proclaimed 'No Riders Allowed'. However, for sheer drama, the little colony of Shay engines made a close rival. They looked both comical and frightening; thick wooden buffer beams, cylinders whirring a grinding drive shaft, and hideous spark-arresting chimneys, which, I am pleased to say, were totally ineffectual! The Shays were blessed with unequalled de-

Plate 88
Dawn in the Insular Lumber Company's yard at Fabrica finds 70-year-old 2-truck Shay No. 1 getting up steam for a day's work around the sawmills

Plate 89
Following pages: The sugar milling season in the Philippines is a 24-hour day, seven-day week operation. This picture, made in the plantation in the early hours of the morning, shows again the wonderful ex Bacalod-Murcia Railway No. 5, a 3 ft. gauge Alco 2–6–0 of 1924. Notice the bale of bagasse on the engine's running plate, and also her enormous yellow tender taken from an old Baldwin 0–6–2

crepitude; never had I seen or imagined anything like it before.

The presence which emanated from Mallet No. 7 was awesome, and on my first evening at Fabrica I witnessed her arrive with a big train of freshly cut logs from the mountain. Flinging sparks and rasping steam off her cylinders, she rolled by with the long rake of log cars eerily sliding after her into the gloom. She stopped at the sawmill on a ledge high above a huge artificial lake. Here the logs were to be mechanically pushed from the waggons and allowed to crash violently down the slope into the lake below, from which a conveyor took them up into the sawmill. From the opposite side of the lake I watched No. 7 ease the leading log car into position and after grappling for a few seconds, the pusher dislodged an enormous trunk and sent it crashing down into the pan. The ground shook under the impetus and when the trunk finally hit the lake it sent a terrific spray of water some forty feet into the air. With three trunks sent down in rapid succession, No. 7 eased her waggons forward emitting grotesque sounds from her four leading cylinders, all of which had their valves completely out of alignment. Like an anguished creature she moved forwards and, backlit by sodium lights from the sawmill, her acrid silhouette was seen to be ever changing in ghostly patterns amid fire and swirling steam. When the logs from each waggon had been pushed into the pan, No. 7 eased her empties back up the gradient away from the mill. Each time she slipped, a flourish of sparks swirled into the air and steam spurted from innumerable places accompanied by the most hideous sounds ever emitted from a locomotive; it is no exaggeration to say that she sounded as if she were possessed by all the devils in hell. Small wonder the Company had reduced her boiler pressure, and this, combined with her unco-ordinated cylinders, had sapped much of her former power; No. 7 had long since become a ghost which had refused to die.

Later that night after the Mallet had gone back up the mountain, Lima Shay No. 10 (plate 17), began tripping between the sawmill and planing mill, a distance of about one mile. The whole town knew she was out, on account of the incredible racket she made. The fire effects were a spectacle within themselves, with bits of flaming teak being flung up to thirty yards from the trackside: the engine's bizarre shape consummated the effect to an unimaginable level.

Next day, up in the mountain, three-truck Shay No. 12 had arrived at Maaslud with a rake of teak from the cutting area and was waiting to hand it over to No. 7 for the journey down to the mill (plate 14): Dawn broke and No. 12 gently backed the train up to where No. 7 was standing (plate 49), and I set off for the wooden trestle bridge half a mile down the track to catch her on the bridge with a full load, and so enact the final moments of my dream. It was 7.30 a.m., the low angle sunlight was crystal clear and the distant mountain tops were flecked with cloud: it all seemed too good to be true. A chime whistle rang out over the groves and echoed for miles across the still landscape – she was on her way. Snorting her syncopated cacophony, the world's most incredible steam survivor rounded the bend and as she gently eased her way over the viaduct I sensed one of the greatest thrills of my life: the resultant picture is on plate 25; for me the 'portrait of a dinosaur'.

So, in the mid-1970s there exists a line in the best traditions of the old lumber railways of the American Pacific North-west, operated by two evolutionary forms of locomotives which will always be associated with the New World. Long may they survive – for we shall be infinitely the poorer without them.

Plate 90
Opposite: Of typically American utilitarian appearance, a 2-cylinder Baldwin-built 2–10–0 of 1947 trundles along the main lines of Greece under the classification of Λγ. The engine, fired by oil, is heading northwards from Athens

Plate 91
Following pages: Top: An oil-fired Baldwin 2–10–0 of 1947 heads through the springtime landscape of Greece from Athens to Thessaloniki, with an overnight goods train

Plate 92
Bottom: A Finnish Class Tk3 2–8–0 faces a blizzard at Rovaneimi on the Arctic Circle. The engine is about to collect snow ploughs to keep the main lines clear as heavy falls are forecast